Mahmoud E.F. Abdel-Haliem

Genetic Engineering

Mahmoud E.F. Abdel-Haliem

Genetic Engineering

Genetic Engineering

Noor Publishing

Imprint

Cover image: www.ingimage.com

Publisher:
Noor Publishing
is a trademark of
International Book Market Service Ltd., member of OmniScriptum Publishing Group
17 Meldrum Street, Beau Bassin 71504, Mauritius

Printed at: see last page
ISBN: 978-620-0-07339-6

Content page

1. Genetic Engineering 1
  1.1 The development of genetics... 1
  1.2 The technology 3
  1.3 What is gene cloning ? 3
    1.3.1 Preparation of DNA from living cells 5
    1.3.2 What is PCR? 6
      1.3.2.1 The number of applications of PCR 10
      1.3.2.2 Different types of DNA polymerase I enzymes 11
      1.3.2.3 Proofreading 11
    1.3.3 Gene isolation by cloning 12
2. Modification and Restriction enzymes 14
  2.1 DNA modifying enzymes 14
    2.1.1 Nucleases (exonucleases and endonucleases) 14
    2.1.2 Alkaline phosphatase 15
    2.1.3 Polynucleotide kinase 15
    2.1.4 Terminal transferase 15
    2.1.5 Topoisomerases 16
  2.2 Restriction endonucleases enzymes 16
    2.2.1 Type II restriction endonucleases 18
      2.2.1.1 The frequency of recognition sequences in a DNA molecule 19
      2.2.1.2 Blunt ends and sticky ends 19
  2.3 DNA ligase 21
  2.4 Putting sticky ends onto a blunt-ended molecules 24
    2.4.1 Linkers 24
    2.4. 2 Adaptors 24
  2.5 Polymerase 25
3. Transformation (introduction of DNA into living cells) 27
  3.1 Preparation of competent E. coli cells 28
  3.2 Selection of transformed cells 29
  3.3 Cloning Vectors (Vehicles) for gene cloning: plasmids, bacteriophages and bacterial artificial chromosomes 29
  3.3.1 Cloning vectors based on E. coli plasmids 30
    3.3.1.1 Basic features of plasmids 30
    3.3.1.2 Size and copy number 32
    3.3.1.3 Conjugation and compatibility 34
    3.3.1.4 Plasmid classification 34
  3.3.2 Cloning vectors based on M13 bacteriophage 35
  3.3.3 Cloning vectors based on Lambda (λ) bacteriophage 35
  3.3.4 Cloning vector based on Cosmid 37
  3.3.5 Cloning vectors based on YAC (yeast artificial chromosome)............ 37
  3.3.5.1 Essential components of YAC vectors 38
4. DNA Sequence 39
  4.1 Restriction digest DNA in the laboratory..................................... 39
  4.2 Separation of molecules by gel electrophoresis 40
  4.3 Visualizing DNA molecules in a gel by staining 41
    4.3.1 Autoradiography of radioactively labelled DNA 42
    4.3.2 Sanger dideoxy sequencing method (the chain termination) 44
    4.3.3 Maxam and Gilbert sequencing Technique (Chemical cleavage)...... 47

4.4 Sequencing PCR products.... 49
4.4.1 Direct sequencing of PCR products 49
4.4.2 Thermal cycle sequencing.... 49
4.5 Computer aided sequence analysis 50
4.6. Edmann degradation for protein 51
5. What is bioinformatics? 54
5.1 Bioinformatics at the heart of a scientific revolution 54
5.2 Bioinformatics are three things 55
5.3 To understand a genome, three distinct types of analysis must be carried out 55
5.4 What is Bioinformatics used for? 55
5.5 Exponential growth of data banks 56
5.5.1 NCBI (National Center for Biotechnology Information) 57
5.5.1 NCBI (National Center for Biotechnology Information) 57
5.5.1.1 The NCBI web entry site provides a surface to enter various Databases 57
5.5.2 Nucleotide databases 58
5.5.2.1 GenBank 58
5.5.2.2 dbEST.... 59
5.5.2.3 dbGSS 59
5.5.2.4 dbSTS 60
5.5.2.5 dbSNP 60
5.5.3 GEO: Gene Expression Omnibus 61
5.5.3.1 How to retrieve a database entry via the NCBI entry site ? 61
5.5.3.2 The database entry contains additional links 61
5.5.3.2.1 Link organism 61
5.5.3.2.2 Link MEDLINE 62
5.5.3.2.2.1 Link mRNA 62
5.5.3.2.2.2 Link CDS 62
5.5.3.2.2.3 Link protein id 62
5.5.4 Searching by subject search 62
5.5.5 Searching by authors 63
5.5.6 Truncations 63
5.5.7 Boolean Operators 64
5.5.8 How to proceed when too many entries are identified by your search term ? 64
5.5.9 Combining previous searching 64
5.5.10 BLAST 65
5.5.10.1 Nucleotide BLAST 65
5.5.10.2 The BLAST Search box 66
5.5.10.2.1 Nucleotide Sequence Databases that can be used for searches 66
5.5.10.2.2 How to perform a BLAST search ? 67
5.5.10.2.3 Pairwise BLAST 68
5.5.11 Protein Sequence Databases 69
5.5.11.1 The SWISS-PROT Database 69
5.5.11.2 Nucleotide Sequence Database 70
5.5.12 Human Proteomics Initiative 71
5.5.13 Related services 72
5.5.14 The InterPro Database 72

5.5.14.1 InterPro Member Databases 73
5.5.15 BCM Search Launcher (Multiple Sequence Alignments) 74
5.5.15.1 Sequence analysis of the coding regions of both AtPIP5K-f and AtPIP5K-x 74
5.5.16 The MIPS Database 76
5.5.16.1 The MAtDB web page 77
5.5. 16.2 The Arabidopsis Information Resource (tair) 78
5.5.16.3 The TIGR Database 79
5.5.16.4 Enzyme cutter 79
5.5.16.5 Promoter analysis 79
5.5.17 Program Rice Genome Research (PRGR) 80
5.5.18 Rice membrane proteins library (RMPL) 80
5.5.19 Metabolic / Enzyme databases 80
5.5.19.1 BRENDA: Enzyme Information Service 80
5.5.20 KEGG: Kyoto Encyclopedia of Genes and Genomes 80
5.5.20.1 EMP: Enzymes and metabolic pathways 81
5.5.20.2 WIT: What is there? 81
5.5.20.2.1 Interactive Metabolic Reconstruction on the Web 81
5.5.20.2.2 Boehringer Mannheim-Biochemical Pathways 81
5.5.21 Yeast Pathways in MIPS 82
5.5.22 Human database 82
5.5.22.1 The Genome Database (GDB) 82
5.5.22.2 GeneCards 82
5.5.22.3 HUGE: Database of Human Unidentified Gene-Encoded Large Proteins 83
5.5.22.4 Mitomap: A Human Mitochondrial Genome Database 83
5.5.22.5 The Whole Mouse Catalog 83
5.5.22.6 Bovine Genome Database 83
5.5.22.7 The WWW Virtual Library: Model Organisms 84
5.5.22.8 How to submission of sequence in GenBank? 84
5.5.22.9 How to draw the family tree? 84
6.Transgenic Organisms 85
6.1 Hybridization probe (random prime labelling system) 85
6.2 Blotting (RNA or DNA) 85
6.3 Northern blotting 86
6.4 DNA preparation (genomic DNA from plant tissue) 86
6.5 Southern blot 87
6.6 Protein extractions from plant leaves and seeds 88
6.7 SDS-polyacrylamide gel electrophoresis of proteins (PAGE) 89
6.8 Protocol of Western blot 89
6.9 Western blot 89
6.10 Immune-Blot assay 90
6.11 Transgenic Organisms 91
6.11.1Transgenic animals 92
6.11.2 Transgenic plants (Plant gene transfer) 92
References 95

1. Genetic Engineering

Genetic engineering represents a fusion of pure science and economics, of market and laboratory. The key of genetic engineering and secret of all heredity is DNA. This molecule, found within every living cell and controls the function and development of all life on earth.

1.1 The development of genetics

Gregor Mendel formulated a set of rules to explain the inheritance of biological characteristics. The basic assumption of these rules is that each heritable property of an organism is controlled by a factor, called a **gene**, that is a physical particle present somewhere in the cell. The rediscovery of Mendel`s laws in 1900 marks the birth of genetics, the science aimed at understanding what these genes are and exactly how they work. For the first thirty years of its life this new science grew at an astonishing rate. The idea that genes reside on chromosomes was proposed by W. Sutton in 1903, and received experimental backing from Thomas Hunt Morgan in 1910. Morgan and his colleagues at Columbia University in New York developed the techniques for gene mapping, and by 1922 had produced a comprehensive analysis of the relative positions of over 2000 genes on the four chromosomes of the fruit fly, *Drosophila melanogaster*.

Despite the brilliance of these classical genetic studies, there was no real understanding of the molecular nature of the gene until the 1940s. Indeed, it was not until the experiments of Avery, McCLeod and McCarty in 1944, and of Hershey and Chase in 1952, that anyone believed DNA to be the genetic material: Up until then it was widely thought that genes were made of protein. The discovery of the role of DNA was a tremendous stimulus to genetic research, and many famous biologists (Delbrück, Chargaff, Crick and Monod were among the most influential) contributed to the second great age of genetics.

In the 14 years between 1952 and 1966 the structure of DNA was elucidated, the genetic code cracked, and the processes of transcription and translation desribed. Then in the years 1971-1973 genetic research was thrown back in to gear by what at the time was described as a revolution in experimental biology. A whole new methodology was developed, enabling previously impossible experiments to be planned and carried out, if not was ease, then at least with success. These methods, referred to as **recombinant DNA technology** or **genetic engineering**, and having at their core the process of **gene cloning**, sparked another great age of genetics. They led to rapid and efficient DNA sequencing techniques that enabled the structures of individual genes to be determined, reaching a culmination in the 1990 with the massive genome sequencing projects, including the human project which was completed in 2000. They led to procedures for studying the regulation of individual genes, which have allowed molecular biologists to understand how aberrations in gene regulation can result in human diseases such as cancer. The techniques for DNA cloning paved the way to the modern fields of **biotechnology (genomics** and **proteomics**), which puts genes to work the production of proteins and other compounds needed in basic research, agriculture as: Diseases and herbicide resistance have been introduced into crops (e.g. Basta resistance in cereals, viral diseases in tomato and cucumber, bacterial diseases in potato and tomato), yield and quality improvement (e.g. potato, wheat, maize, soybean and clover), tolerance of environment extremes in crops (drought, cold, salinity), medicine as manufacture of pharmaceutical products in plants, industrial processes as biofuels is being developed from genetically engineered oilseed rape and many other fields.

1.2 The technology

Starting with the double helix of Watson and Crick in 1953, knowledge of DNA has increased exponentially. Key discoveries include the genetic code, how protein synthesis takes place, what gene is in chemical terms, and how to determine the sequence of DNA (Table 1-1). At all of these stages, biochemical knowledge of enzymes, cofactor requirements and generally of how to do things in the test tube (*in vitro*) has also increased. This has culminated in the technology by which DNA may be manipulated, sequenced, cut, identified, duplicated and transferred in and out of cells and organisms.

Table 1-1 Some keys dates in molecular Biology since the Watson and Crick discovered

1953	Watson and Crick publish their paper which proposes a double helical structure for DNA with the bases on the inside of the helix, matching A: T, C: G
1957	DNA polymerase discovered by Korberg
1962	Restriction enzymes discovered
1966	The genetic code deciphered
1972-3	Cloning in bacteria first carried out
1975	Southern blotting invented
1975	Sanger publishes his method for DNA sequencing
1981	First transgenic animals
1985	Polymerase chain reaction invented

1.3 What is gene cloning?

A clone is an identical copy. This term originally applied to cells of a single type, isolated and allowed to reproduce to create a population of indentical cells. **DNA cloning** involves separating a specific gene or DNA segment from a larger chromosome, attaching it to a small molecule of carrier DNA, and then replicating this modified DNA thousands or millions of times through both an increase in cell number and the creation of multiple copies of the cloned DNA in each cell. The result is selective amplification of a particular gene or DNA segment. The basic steps in a gene cloning experiment are as follows (Fig.1.1):.

1) A fragment of DNA was Cutting at precise locations. Sequence-specific endonuclease (restriction endonucleases) provide the necessary molecular scissors.

2) Selecting a small molecule of DNA capable of self-replication. These DNAs are called cloning vectors (a vector is a delivery agent). They are typically plasmids or viral DNAs. Vector acts as a **vehicle** that transports the gene into a host cell (vector must be able to replicate, must be some means of detecting its presence and must be some way to introduce vector DNA into a cell).

3) Joining two DNA fragments covalently. The enzyme DNA ligase links the cloning vector and DNA to be cloned. Composite DNA molecules comprising covalently linked segments from two or more sources are called **chimera** or **recombinant DNAs**.

4) Moving recombinant DNA from the test tube to a host cell which is usually a bacterium, although other types of living cell can be used, within the host cell the vector multiples, producing numerous identical copies not only of itself but also of the gene that it carries.

5) Selecting or identifying host cells that contain recombinant DNA and when the host cell divides, copies of recombinant DNA molecule are passed to the progeny and further vector replication takes place. After a large number of cell divisions, a colony, or clone of identical host cells is produced. Each cell in the clone contains one or more copies of recombinant DNA molecule.

The methods used to accomplish these and related tasks are collectively referred to as **recombinant DNA technology** or, more informally, **genetic engineering**.

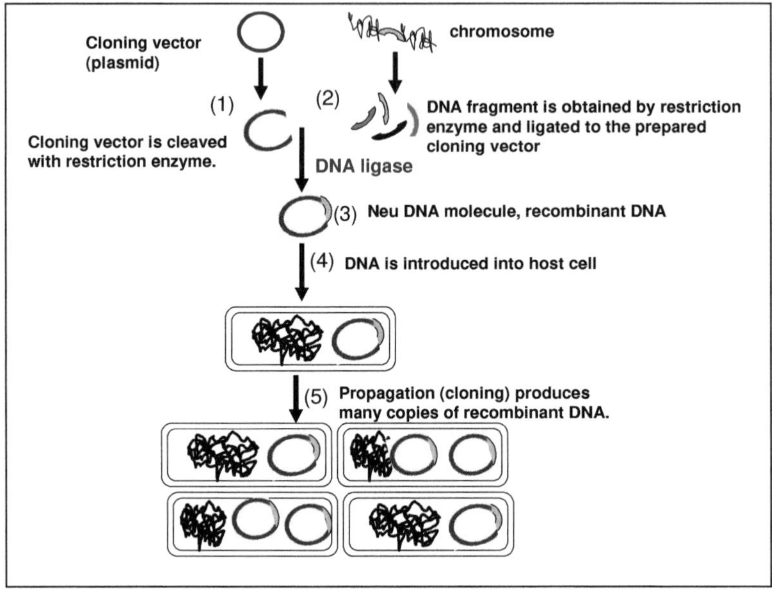

Fig. 1.1 Schematic illustration of DNA cloning. A cloning vector and eukaryotic chromosomes are separately cleaved with the same restriction endonuclease. The fragments to be cloned are then ligated to the cloning vector. The resulting recombinant DNA is introduced into a host cell where it can be cloned (propagated).

1.3.1 Preparation of DNA from living cells

In genetic engineer need to prepare at least two kinds of DNA. Firstly, total cell DNA will often be required as a source of material from which to obtain genes to be cloned. Total cell DNA may be from a culture of bacteria, from a plant, from animal cells, or from any other type of organism. The second type of DNA that will be required is pure plasmid DNA. Preparation of plasmid DNA from a culture of bacteria follows the same basic steps as purification of total cell DNA, with the difference that at some stage the plasmid DNA must be separated from the main bulk of chromosomal DNA also present in the cell, or phage DNA will be needed if a phage cloning vehicle is to be used. Phage DNA is generally prepared from bacteriophage particles rather than from infected cells, so there is no problem with contaminating bacterial DNA.

1.3.2 What is PCR?

The polymerase chain reaction (PCR) is a rapid procedure for *in vitro* enzymatic amplification of a specific DNA segment, like molecular cloning. If we know the sequence of at least the flanking parts of a DNA fragment to be cloned, we can hugely amplify the number of copies of that DNA fragment, using the polymerase chain reaction, a process conceived by Kary Mullis in 1983. The PCR procedure is carried out in a single test tube simply by mixing three nucleic acid segments: the segment of double-stranded DNA to be amplified and two single-stranded oligonucleotide primers flanking it. Additionally, there is a protein component (a DNA polymerase), appropriate deoxyribonucleotide triphosphate (dNTPs), a buffer, salts and placing the tube in a thermal cycler, a piece of equipment that enables the mixture to be incubated at a series of temperatures that are varied in a programmed manner.

1-The mixture is heated to 94 °C (denaturation temperature), at which temperature the hydrogen bonds that hold together the two strands of the double-stranded DNA molecule are broken and release single strand DNA to act as template.

2- The mixture is cooled down to 50-60 °C (annealing temperature) in the presence of 5`and 3` primers, $MgCl_2$ and dNTPs, The two strands of each could join back together at this temperature, but most do not because the mixture contain large excess of short DNA molecules, called synthetic oligonucleotides or primers, which anneal or attach to the DNA template molecules at specific positions. The oligonucleotides serve as replication primers that can be extended by DNA polymerase.

3- The temperature is raised to 72°C (the extension temperature). This is the optimum working temperature for heat stable DNA polymerase I, such as *Taq* polymerase (derived from a bacterium *Thermus aquaticus*, that lives in hot springs at 90°C, this enzyme is thermostable, meaning that they resistant to denaturation by heat treatment or which remains active after every heating steps and does not have to be replenished) in the mixture, and attaches to the end of each primer and synthesizes new strands of DNA, complementary to the template DNA molecules, during this step of the PCR.

4- The temperature is increased back to 94°C. The double-stranded DNA molecules, each of which consists of one strand of the original molecule and one new strand of DNA, denature into single strands. This begins a second cycle of denaturation (heating), annealing (cooling) and synthesis (replication), at the end of which there are eight DNA strands. By repeating the cycle 25-36 times the double stranded molecule that we began with is converted into over 50 million new double-stranded molecules, about 1 picogram of DNA into micrograms, sufficient for most analytical and for many other uses.

At the end of a PCR a sample of the reaction mixture is usually analysed by agarose gel electrophoresis sufficient DNA have been produced for the amplified fragment to be visible as a discrete band after staining with ethidium bromide. The two strands of DNA are only linked by non-covalent forces, they can easily be separated in the laboratory, for example by increased temperature or high pH. Separation of the two strands, a process known as denaturation, is readily reversible. Reducing the temperature, or the pH, will allow hydrogen bonds between complementary DNA sequences to reform; this is referred to as re-annealing. If DNA molecules from different sources are denatured, then mixed and allowed to re-anneal, it is possible to form hydrogen bonds between regions of DNA that are similar (but not necessarily identical).

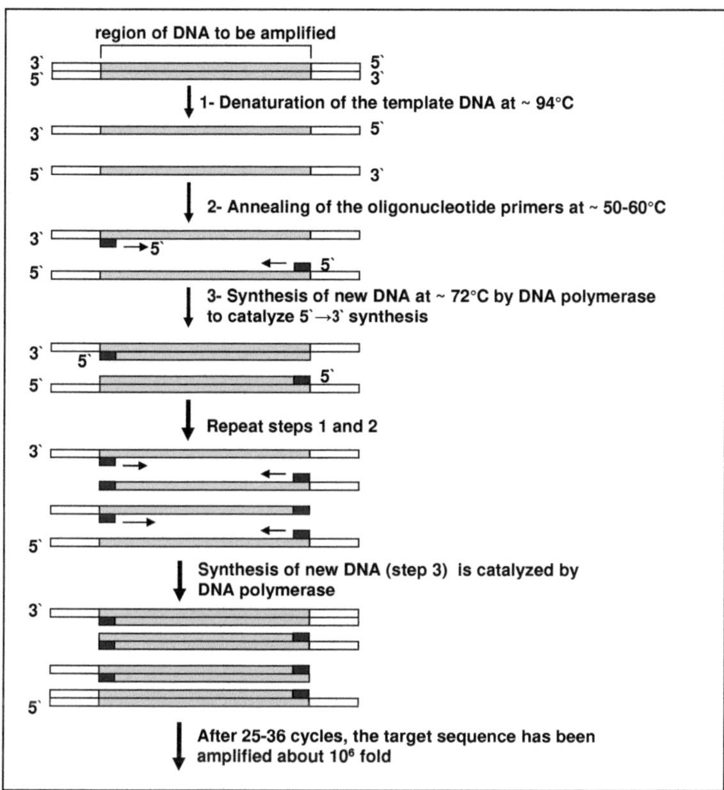

Fig. 1.2 Amplification of a DNA segment by the polymerase chain reaction. The PCR procedure has three steps. DNA strands are (1) denaturated by heating (2) annealed to an excess of short synthetic DNA primers that flank the region to be amplified; (3) new DNA is synthesized by polymerization. The three steps are repeated for 25 or 36 cycles. The thermostable DNA polymerase *Taq* (from *Thermus aquaticus*, a bacterial species that grows in hot springs) is not denaturated by heating steps.

Hybridization between single-stranded DNA molecules forms the basis of the use of DNA probes to detect specific sequences in samples of DNA. The specificity of the reaction can be adjusted by altering the temperature and / or the ionic strength. Higher temperature, or lower ionic strength, gives greater stringency of hybridization, i.e., the probe has to be more like the sequence to be detected. High-stringency hybridization is used to detect closely related sequences, or to distinguish between sequences with only small differences, while low stringency conditions are used to detect sequences that are only remotely related to your probe. Note that there are three hydrogen bonds linking guanine and cytosine while the adenine-thymine pairing has only two hydrogen bonds. The two DNA strands are therefore more strongly attached in those regions with a high G + C content. Because of this, such regions are more resistant to denaturation and conversely reanneal more readily. The influence of base composition on the ease of separation of two nucleic acid strands may play an important role in the control of processes such as the initiation of RNA synthesis where an A-T rich region may facilitate the initial separation of the DNA strands.

During each cycle of a PCR the reaction mixture is transferred between three temperatures. The annealing temperature can affect the specificity of the reaction. If the temperature is too high no hybridization take places, instead the primers and templates remain dissociated. And also, if the temperature is too low, mismatched hybrids not all the correct base pairs have formed. The ideal annealing temperature must be low enough to enable hybridization between primer and template, but high enough to prevent mismatched hybrids forms. This temperature can be estimated by determining the **melting temperature** or **T_m** of the primer-template hybrid. The T_m is the temperature at which the correctly base-paired hybrid dissociates (melts); a temperature 1-2 °C below this should be low enough to allow correct primer-template hybrid to form, but too high for a hybrid with a single mismatch to be stable.

$$T_m = 4°C \times (G + C) + 2°C \times (A + T)$$

Where: (G + C) is the number of G an C nucleotides in the primer sequence, and (A + T) is the number of A and T nucleotides. The annealing temperature for a PCR experiment is

Example: Calculating the Tm of the primer.

Primer sequence: 5`-ACTTGGAATCGGTAACGTATG-3´

$T_m = 4°C \times (G + C) + 2°C \times (A + T)$

$T_m = 4°C \times (6+ 3) + 2°C \times (6+6) = 36 + 24 = 60$ °C.

The annealing temperature is 58 °C in this example.

determined by calculating the T_m for each primer and using a temperature 1-2 °C below the calculation.

A PCR experiment can be completed in a few hours, whereas it takes weeks if not months to obtain a gene by cloning. Why then is gene cloning still used? This is because of **two limitations with PCR:**

1- In order for the primers to anneal to the correct positions, either side of the gene of interest, the sequences of these annealing sites must be known. It is easy to synthesize a primer with a predetermined sequence, but if the sequences of the annealing sites are unknown then the appropriate primers cannot be made. This means that PCR can not be used to isolate genes that have not been studied before, that has to be done by cloning.

2- There is a limit to the length of DNA sequence that can be copied by PCR. Four kilobases (kb) can be copied fairly easily, and segments up to 40 kb can be dealt with using specialized techniques, but this shorter than the lengths of many genes, especially those of humans and other vertebrates.

1.3.2.1 The number of applications of PCR

It seems infinite and is still growing. They include direct cloning from genomic DNA or cDNA, *in vitro* mutagenesis and engineering of DNA, genetic fingerprinting of forensic samples, assays for the presence of infectious agents, prenatal diagnosis of genetic diseases, analysis of allelic sequence variations, analysis of RNA transcript structure, genomic footprinting, and direct nucleotide sequencing of genomic DNA and cDNA. The PCR method is also important in advancing the goal of whole genome sequencing. For example, the mapping of expressed sequence tags to particular chromosomes often involves amplification of the EST (expressed sequence tag) by PCR, followed by hybridization of the amplified DNA to clones in an ordered library. Also the PCR technology is highly sensitive, so can detect and amplify as little as one DNA molecule in almost any type of sample. Although DNA degrades over time, PCR has allowed successful cloning of DNA from samples more than 40,000 years old. Investigators have used the technique to clone DNA fragments from the mummified remains of humans and extinct animals such as the woolly mammoth, creating the new fields of molecular archaeology and molecular paleontology.

1.3.2.2 Different types of DNA polymerase I enzymes:

1. *Taq* polymerase: Add A or T at the end, lack a proofreading activity and as a result is unable to correct its errors. So is not accurate copy of the template copy and active at 72 °C.

2. *Pfu* polymerase: Blunt end, has proofreading, active at 68 °C.

3. Hf_2 polymerase: Add A or T at the end, has proofreading, active at 72 °C.

All nucleic acids are synthesized in the 5` to 3` direction. That is, they are elongated by successive addition of nucleotides to the free 3` OH group of the preceding nucleotide. The phosphate to make the link is provided by the substrate, which is the nucleoside 5`-triphosphate; the energy required is provided by the release of the two terminal phosphate groups.

1.3.2.3 Proofreading

Each of *E. coli* DNA polymerase possesses an exonuclease activity in the 3`to 5´ direction, the reverse to the synthesis direction. This means that, if a nucleotide is inserted incorrectly during DNA synthesis, the enzyme is able to retrace its steps, removing the mismatch and replacing it with the correct nucleotide (Fig.1.3). This is called the proofreading function and helps to ensure that DNA replication is virtually error-free.

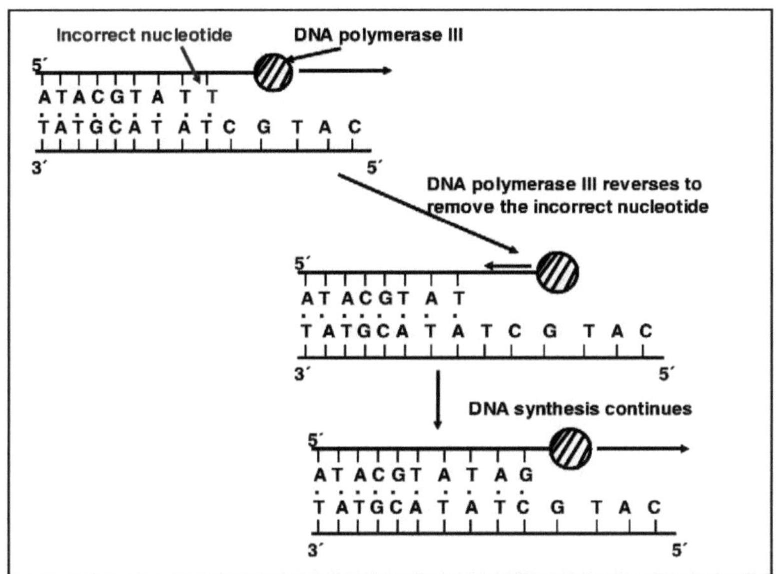

Fig. 1.3 Elimination of mispaired bases by proof-reading. An incorrect base has been added to the growing DNA strand; this will prevent further extension. The mispaired base is removed by the 3`-5` exonuclease activity of the DNA polymerase. The correct base is added to the 3`end of the growing strand; DNA synthesis continues.

Note: Gene cloning and PCR are relatively straightforward procedures. Why, then, have they assumed such importance in biology? The answer is largely because both techniques can provide a pure sample of an individual gene, separated from all the other genes in the cell.

1.3.3 Gene isolation by cloning

DNA cloning in bacterium *Escherichia coli*, the first organism used for recombinant DNA work and still the most common host cell. *E. coli* has many advantages; its DNA metabolism (like many other of its biochemical processes) is well understood; many naturally occurring cloning vectors associated with *E. coli*, such as plasmids and bacteriophages (bacterial viruses; also called phages), are well characterized; and techniques are available for moving DNA expeditiously from one bacterial cell to another.

Particularly important to recombinant DNA technology is a set of enzymes (Table 1-2) made available through decades of research on nucleic acid metabolism. Two classes of enzymes lie at the heart of the general approach to generating and propagating a recombinant DNA molecule (Fig. 1.1). First, **restriction enzymes** (also called **restriction endonucleases**) recognize and cleave DNA at specific DNA sequences (recognition sequences or restriction sites) to generate a set of smaller fragments. Second, the DNA fragment to be cloned can be joined to a suitable cloning vector by using DNA ligase to link the DNA molecules together. The recombinant vector is then introduced into a host cell, which amplifies the fragment in the course of many generations of cell division.

Table 1.2 Some enzymes used in recombinant DNA technology

Enzymes	Function
Restriction endonuclease (type II)	Cleave DNAs at specific base sequences
DNA T4 ligase	join two fragments of DNA molecules
DNA polymerase I (*E. coli*)	fills gaps in duplexes by addition to 3`ends
Reverse transcriptase	makes a cDNA copy of an RNA molecule
Polynucleotide kinase	adds a phosphate to the 5`-OH end of polynucleotides to permit ligation
Alkaline phosphatase	Removes terminal phosphates from the 5` end of DNA
Terminal transferase	Adds deoxyribonucleotides onto the 3`terminus of a DNA molecule

2. Modification and Restriction enzymes

2.1 DNA modifying enzymes

There are numerous enzymes that modify DNA molecules by addition, removal or change the conformation of DNA.

2.1.1 Nucleases (exonucleases and endonucleases)

Nucleases degrade DNA molecules by breaking the phosphodiester bonds that link one nucleotide. There are two kinds of nuclease; exonucleases remove nucleotides from the end of a DNA molecule or endonucleases are able to breake internal phosphodiester bonds within a DNA molecule (Fig. 2-1).

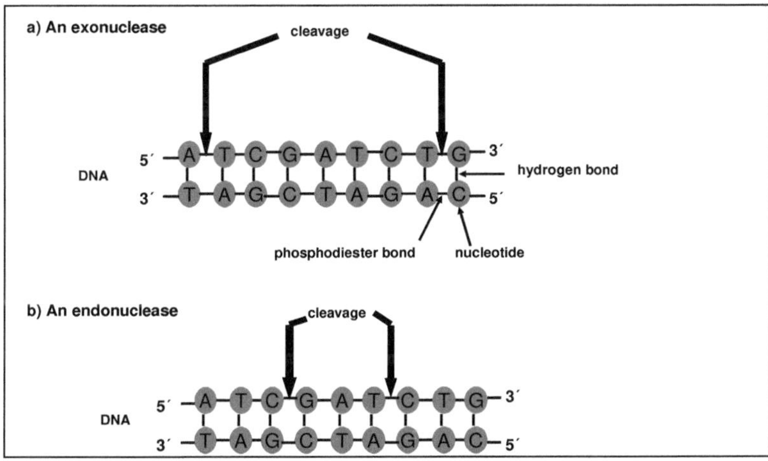

Fig. 2-1 Two different kinds of nuclease. a) An exonuclease, which removes nucleotides from the end of a DNA molecule and **b) an endonuclease**, which breaks internal phosphodiester bonds.

Many types of exonucleases lie in the number of strands that are degraded when a double-stranded molecule is attacked. The enzyme *Bal*31(*Alteromonas espejiana*) is an example of an exonuclease that removes nucleotides from the end of both strands of a double-stranded molecule, but another example *E. coli* exonuclease III degrade just one strand (3´terminus of a double stranded molecule. The enzyme S1 endonuclease (from the fungus *Aspergillus oryza*) only cleaves single strands, whereas deoxyribonuclease I (DNase I), which is prepared from cow pancreas, cuts both single and double stranded molecules, which is not specific in that it attacks DNA at any internal phosphodiester bond and the end result of prolonged DNase I action is therefore a mixture of mononucleotides and very short oligonucleotides.

Only to know, S1 endonuclease has a role in the studying a DNA-RNA hybrid. This enzyme degrades single-stranded DNA or RNA polynucleotides, including single-stranded regions, but has not affect on double-stranded DNA or DNA-RNA hybrids. If a DNA molecule containing a gene is hybridized to its RNA transcript, and then treated with S1 nuclease, the non-hybridized single-stranded DNA regions at each end of the hybrid are digested, along with any looped-out introns (Fig. 2-2). The result is a completely double-stranded hybrid. The single-stranded DNA fragments protected from S1 nuclease digestion can be recovered if the RNA strand is digested by treatment with alkali.

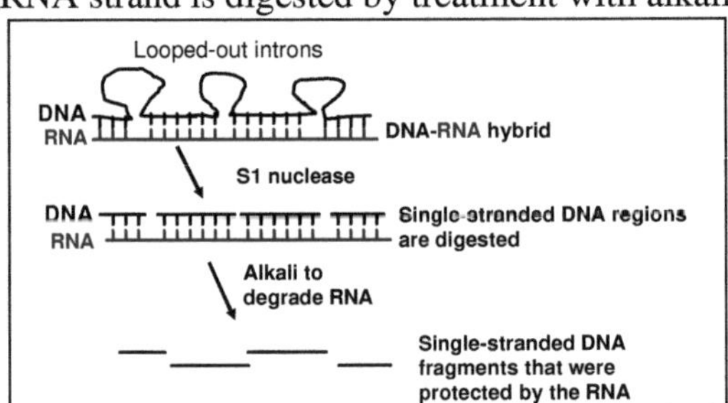

Fig. 2-2 The effect of S1 nuclease on a DNA-RNA hybrid.

2.1.2 Alkaline phosphatase

Alkaline phosphatase purified from calf intestinal tissue or *E. coli*, which removes the phosphate group present at the 5` termini of a DNA molecule (Fig. 2-3a)

2.1.3 Polynucleotide kinase

This enzyme purified from *E. coli* infected with T4 phage, which adds phosphate groups onto free 5` termini of a DNA molecule (Fig. 2-3b).

2.1.4 Terminal transferase

This enzyme purified from calf thymus tissue, which adds one or more deoxyribonucleotides onto the 3`terminus of a DNA molecule (Fig. 2-3c)

2.1.5 Topoisomerases

The enzymes are able to change DNA from supercoil to the conformation of covalently closed-circular DNA. If one of the polynucleotide strands is broken the double helix reverts to the normal relaxed state (open circular). By introducing or removing supercoils (DNA replication) and also most plasmids exist in the cell as supercoiled molecules configurations, so important in genetic engineering (Fig. 2-3d).

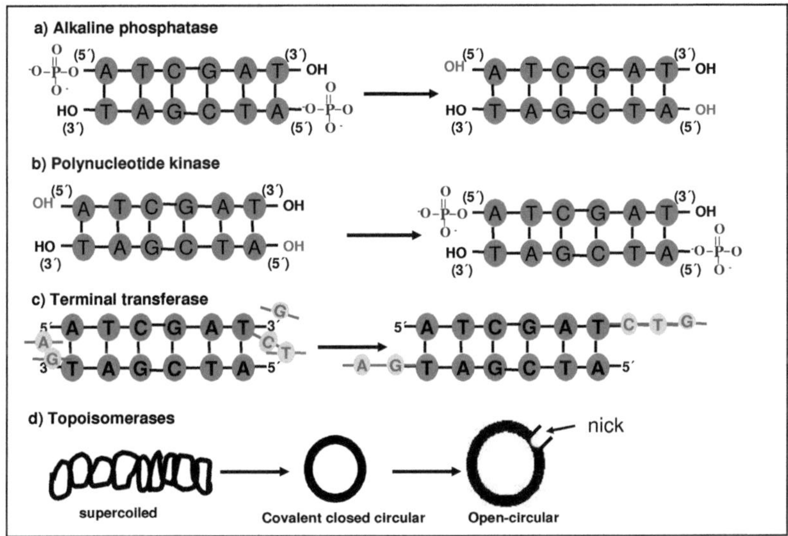

Fig. 2-3 The reactions catalysed by DNA modifying enzymes. a) alkaline phosphatase, which removes 5`-phosphate groups. **b) polynucleotide kinase,** which adds 5`-phosphate groups. **c) terminal transferase,** which adds deoxyribonucleotides to the 3´-termini of polynucleotides. **d) topoisomerases,** the double helix of the plasmid DNA is partially unwound during the plasmid replication process by topisomerases enzymes, and conformation of circular double stranded DNA as supercoiled (which both polynucleotide strands are intact, and the more technical name of covalent closed circular DNA, if one of the polynucleotide strands is broken the double helix and called open circulr.

2.2 Restriction endonucleases enzymes

Restriction enzymes recognize specific base sequences in double-helical DNA and cleave both strands of the duplex at specific places and cause cleavage (hydrolysis of phosphodiester bonds) in or near that specific region. They are indispensable for analyzing chromosome structure, sequencing very long DNA molecules, isolating genes, and creating new DNA molecules that can be cloned. These enzymes are found **only** in a wide range of prokaryotes (bacterial species), **not** in human, animal or plant, and cut **only** DNA **not** RNA. Many viral, bacterial, animal, or plant DNA molecules are substrates for the enzymes.

Restriction enzymes cut or cleavage a recognize specific base sequences, and more 3000 differnt enzymes. Werner Arber discovered in early 1960s that their biological function is to recognize and cleave foreign DNA (the DNA of an infecting virus, for example); such DNA is said to be restricted. In the host cell´s DNA, the sequence that would be recognized by its own restriction endonuclease is protected from digestion by methylation of the DNA. Many restriction enzymes recognize specific sequences of four to eight base pairs and hydrolyze a phosphodiester bond in each strand in this region. The recognized sequence is palindromic and cleavage sites are symmetrically positioned. DNA can be methylated in one strand or two strands. Restriction enzymes cleavage the specific base of DNA without methylated. The methylation protected DNA from digestion and degradation (Fig. 2-4).

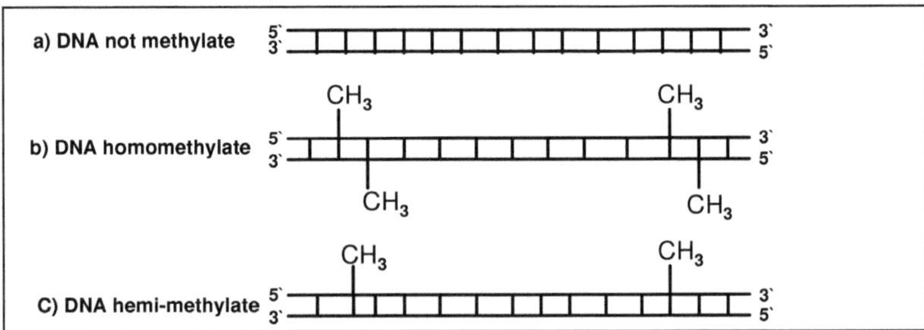

Fig. 2-4 Different DNA metylated forms. A) DNA not metylated, which DNA is not attack with methyl group. **b) DNA homomethylated,** which DNA is attacked with methyl group on the both strand (which is not digest with restriction enzymes). **c) DNA hemimethylated,** which DNA is attack with methyl group only on one strand (only one strand can digest with restriction enzymes which is not methylated).

Over thousands of restriction endonucleases have been discovered in different bacterial species, and more than 100 different DNA sequences are recognized by one or more of these enzymes. There are three types of restriction endonucleases, designated I, II, and III (Table 2-1). Types I and III are generally large, multisubunit complexes containing both the endonuclease and methylase activities, and a limit role in genetic engineering. Type II restriction endonucleases are cutting enzymes that are so important in gene cloning. To check the restriction analysis (site map) by using rebase (http://rebase.neb.com/rebase).

Table 2-1 Comparison between three types of restriction enzymes

	II	III	I
Protein structure	Restriction enzymes and methylase separated, simple no require ATP	Bifunctional, require the energy of ATP	Bifunctional, require ATP
DNA recognize position	4-8 bp, long symmetrical, 4 bp in the case of *Sau*3A (*Staphylococcus aureus*), 6 bp in the case *Hind*III, *Eco*RI, *Bam*HI and 8 bp in the case of *Not*I	5-7 bp, long asymmetrical (random site)	cleaved in two asymmetrical (random site) TGANN.....variable.....NN TGCTACT
Splitting position	The same recognition positions Cleavage site recognition position	24-26 bp from the recognition position Cleavage site recognition position 24-26 bp	~ 1000 bp from the recognition position Cleavage site recognition position ~1000 bp
e.g.	*Hind*III, *Eco*RI, *Bam*HI	*Eco*P15, *Hinf* (*Haemophilus influenzae*)	*Eco*B, *Eco*K

2.2.1 Type II restriction endonucleases

Type II restriction endonucleases is that each enzyme has a specific recognition sequence at which it cuts a DNA molecule and first isolated by Hamilton Smith in 1970, are simpler, require no ATP, and cleave the DNA within the recognition sequence itself (4-8 bp, symmetrical) as 4 bp in the case of *Sau*3A

(*Staphylococcus aureus*), 6 bp in the case *Hind*III, *Eco*RI, *Bam*HI and 8 bp in the case of *Not*I (Table 2-2),. The extraordinary utility of this group of restriction endonucleases was demonstrated by Daniel Nathans, who first used them to dvelop novel methods for mapping and analyzing genes and genomes.

2.2.1.1 The frequency of recognition sequences in a DNA molecule

The average size of the DNA fragments produced by cleaving genomic DNA with a restriction endonuclease depends on the frequency with which a particular restriction site occurs in the DNA molecule; this in turn depends lagely on the size of the recognition sequence and can be calculated mathematically. In a DNA molecule with a random sequence in which all four molecules with a random sequence in which all four nucleotides were equally abundant should occur once every $4^4 = 256$ bp, and a 6 bp sequence recognized by a restriction endonuclease such as *Bam*HI (GGATCC) would occur on average once every $4^6 = 4096$ bp. These calculations assume that the base pairs are ordered in a random fashion and that the four different base pairs are present in equal proportions (i.e. the DNA had a 50 % C≡G content). Enzymes that recognize a 4 bp sequence would produce smaller DNA fragments from a random sequence DNA molecule; a recognition sequences tend to occur less frequently than this because nucleotide sequences in DNA are not random and the four nucleotides are not equally abundant. In practice, the average size of the fragments produced by restriction endonuclease cleavage of a large DNA can be increased by simply terminating the reaction before completion; the result is called a partial digest.

2.2.1.2 Blunt ends and sticky ends

Some restriction endonucleases cleave both strands of DNA at the opposing phosphodiester bonds, leaving no unpaired bases on the ends, cut in the middle of recognition sequence, often called **blunt ends** (flush end), e.g. *Eco*RV, *Sma*I (Fig. 2-5a). Other restriction endonucleases called **sticky ends** (cohesive ends) cut the two DNA strands in a staggered fashion, leaving short single-stranded (two to four nucleotides) overhangs at each resulting end [5´-overhang (see Fig. 2-5b) or 3`-overhang (see Fig. 2-5c)]. Thus, unpaired strands are referred to as sticky or cohesive ends because base pairing between them can stick the DNA molecule together again.

Table 2-2 Recognition sequences from some type II restriction endonuclease

Enzymes	Organisms	Recognition sequence	Blunt or sticky end, and number of bases
BamH	*Bacillus amyloliquefaciens*	5`-G↓GATCC-3` 3`-CCTAG↑G-5`	5`-sticky end (6)
EcoRI	*Escherichia coli*	5`-G↓AATTC-3 3´-CTTAA↑G-5´	5´-sticky end (6)
HindIII	*Haemophilus influenza*	5`-A↓AGCTT-3 3´-TTACGA↑A-5´	5´-sticky end (6)
NotI	*Nocardia otitidis-caviarum*	5`-GC↓GGCCGC-3 3´-CGCCGG↑CG-5´	5´-sticky end (8)
PstI	*Providencia stuartii*	5`-CTGCA↓G-3´ 3´-G↑ACGTC-5´	3´-sticky end (6)
EcoRv	*Escherichia coli*	5`-GAT↓ATC-3´ 3´-CTA↑TAG-5´	Blunt end (6)
PvuII	*Proteus vulgaris*	5´-GAG↓CTG-3´ 3´- GTC↑GAC-5´	Blunt end (6)
HaeIII	*Haemophilus aegyptius*	5´-GG↓CC-3´ 3´-CC↑GG-5´	Blunt end (4)

Arrows indicate the phosphodiester bonds cleaved by each restriction endonuclease. Note that the name of each enzyme consists of a three-letter abbreviation (in italics) of the bacterial species from which it is derived. Sometimes followed by a strain designation and Roman numerals to distinguish different restriction endonucleases isolated from the same bacterial species. Thus, *Bam*HI is the restriction endonuclease characterized from Bacillus amyloliquefaciens, strain H. Thus, *Hind*III is characterized from Haemophillius influenza, strain d, and thus *Eco*RI is characterized from *Escherichia coli*, strain R.

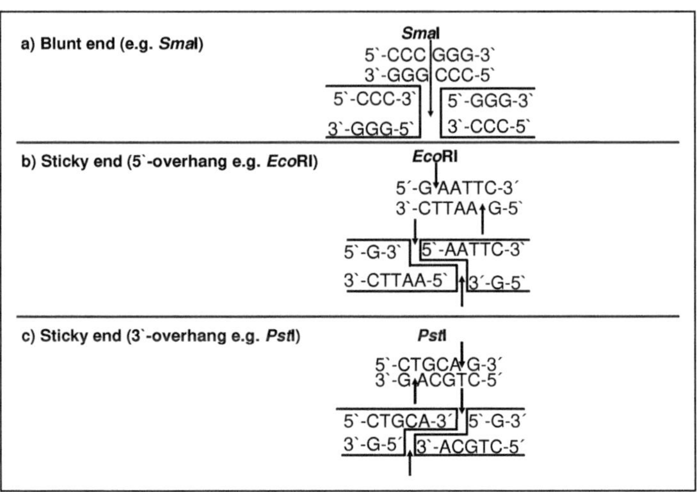

Fig. 2-5 The ends produced by cleavage of DNA with different reaction endonucleases.
a) **blunt end** produced by *Sma*I (*Serratia marcescens*). b) **A sticky 5-end** produced by *Eco*RI. c) **A sticky 3-end** produced by *Pst*I.

2.3 DNA ligase

DNA ligase catalyzes the formation of new phosphodiester bonds between the 3`-OH group at the end of one DNA chain and the 5`-phosphate group at the end of the other (see Fig. 2-6). An energy source is required to drive this endergonic reaction. In *Escherchia coli* and other bacteria, NAD^+ serves this role (cofactor). In eukaryotic and bacteriophage, ATP is the energy source (cofactor). Ligase encoded by T4 bacteriophage. This joining process is essential for the normal synthesis of DNA, the repair of damaged DNA, and the splicing of DNA chains in genetic recombination. DNA ligase can not link two molecules of single-stranded DNA or circularize single-stranded DNA. Rather, ligase seals breaks in double-stranded DNA molecules (Fig. 2-8).

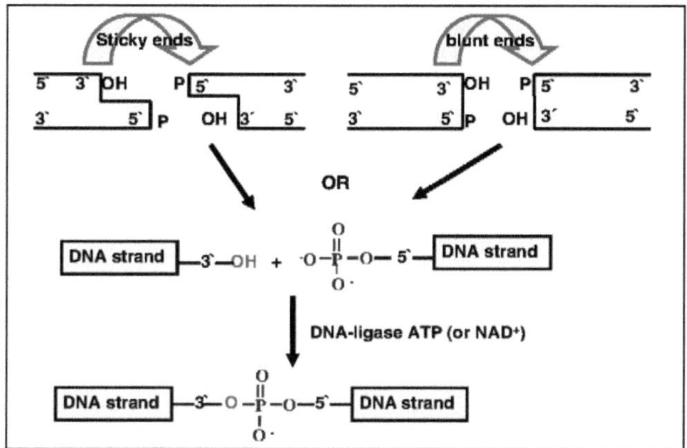

Fig. 2-6 DNA ligase, catalyzes the joining of DNA strands that are part of adouble-helical molecule.

The ligation reaction shows two blunt-ended fragments being joined together. Although this reaction can be carried out in the test tube, it is not efficient. This is because the ligase is unable to "catch hold" of the molecule to be ligated, and has to wait for chance associations to bring the ends together. If possible, blunt end ligation should br performed at high DNA concentrations, to increase the chances of the ends of the molecules coming together in the correct way. In contrast, ligation of complementary sticky ends is much more efficient. This is because compatible sticky ends can base pair with one another by hydrogen bonding (Fig. 2-6), forming a relatively stable structure for the enzyme to work on. If the phosphodiester bonds (dephosphorylation) are not synthesized fairly quickly the sticky ends will fall apart again. These transient, base-paired structure do, however, increase the efficiency of ligation by increasing the length of time the ends are in contact with one another. But the compatible sticky ends are desirable on the DNA molecules to be ligated together in a gene cloning experiment. Often these sticky ends can be provided by digestion both the vector and the DNA to be cloned with the same restriction endonuclease, or with different enzymes (compatible) that produce the same sticky end. In the double helix the two strands are antiparallel, so each end of a double-stranded molecule consists of one 5`-P terminus and one 3`-terminus. Ligation normally takes place between the 5`-P and the 3`-OH ends (Fig. 2.6).

The mechanism of joining in three steps (Fig. 2-7), which was explained by Robert Lehman.

1. ATP (or, in some ligases, NAD^+) donates its activated AMP unit to DNA ligase to form a covalent enzyme-AMP (enzyme-adenylate) complex in which AMP is linked to the ε-amino group of a lysine residue of the enzyme through a phosphoamide bond. Pyrophosphate or nicotinamide mononucleotide, NMN) is concomitantly released.
2. The activated AMP moiety is then transferred from the lysine residue to the phosphate group at the 5`-terminus of a DNA chain, forming a DNA-adenylate complex.
3. The final step is a nucleophilic attack by the 3`-OH group on this activated 5`-phosphorus atom. This sequence of reactions is driven by the hydrolysis of pyrophosphate released in the formation of the enzyme-adenylate complex. Thus, two ~P are spent in constructing a phosphodiester bridge in the DNA backbone when ATP is the adenylate donor. Likewise, two ~P are spent in regenerating NAD^+ from NMN and ATP when NAD^+ is the adenylate donor.

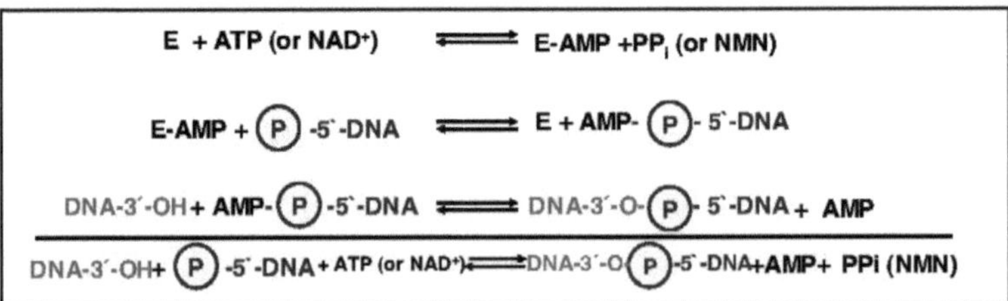

Fig. 2-7. Mechanism of the reaction catalyzed by DNA ligase.

Example for typical DNA liegern (with each other, together, see Fig. 2-8).

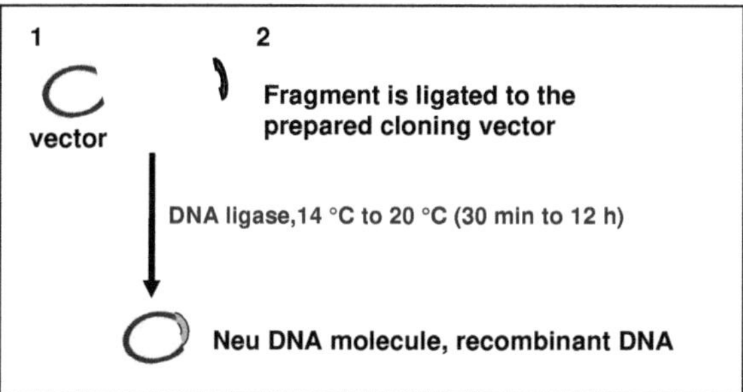

Fig. 2-8 Fragment is ligated in cloning vector.

2.4 Putting sticky ends onto a blunt-ended molecule

A common situation is where the vector molecule has sticky ends, but the DNA fragments to be cloned are blunt-ended. Under these circumstances one of three method can be used to put the correct sticky ends onto the DNA fragments.

2.4.1 Linkers

Blunt ends can also be ligated, albeit less efficiently. Researchers can create new DNA sequences by inserting synthetic DNA fragments (called linkers) between the ends that are being ligated in the case of blunt ends. Inserted DNA fragments with multiple recognition sequences for restriction endonucleases (often useful later as points for inserting additional DNA by cleavage and ligation) are called polylinkers.

2.4. 2 Adaptors

The effectiveness of sticky ends in selectively joining two DNA fragments was apparent in the earliest recombinant DNA experiments. Before restriction endonucleases were widely available, some workers found they could generate sticky ends by the combined action of the bacteriophage λ exonuclease and terminal transferase. The fragments to be joined were given complementary homopolymeric tails. Peter Lobban and Dale Kaiser used this method in 1971 in the first experiments to join naturally occurring DNA fragments. Similar methods were used soon after in the laboratory of Paul Berg to join DNA segments from simian virus 40 (SV40) to DNA derived from bacteriophage λ, thereby creating the first recombinant DNA molecule with DNA segments from different species.

2.5 Polymerase

DNA polymerase are enzymes that synthesize a new strand of DNA complementary to an existing DNA or RNA template. DNA polymerase can function only if the template possesses a double-stranded region that acts as a primer (initiation of polymerization). Three types of DNA polymerase are used in genetic engineering. The first is DNA polymerase I, which is usually prepared from *E. coli* [e.g. *Taq* DNA polymerase purified from *Thermus aquaticus*, used in PCR (polymerase chain reaction)], this enzyme attaches to a short single strand region (primer) or nick region and then synthesizes a completely new strand, degrading the existing strand (Fig. 2-9b). This enzyme has a dual activity of DNA polymerization and degradation.

The second is klenow fragment can still synthesize a complementary DNA strand on a single stranded template, but as it has no nuclease activity it can not continue the synthesis once the nick is filled in (Fig. 2-9c). The final is reverse transcriptase is involved in the replication of virus. This enzyme uses RNA as a template (not DNA) (Fig. 2-9d). The ability of this enzyme to synthesize a DNA strand complementary (cDNA) from an RNA template that is important in cloning of cDNA in genetic engineering.

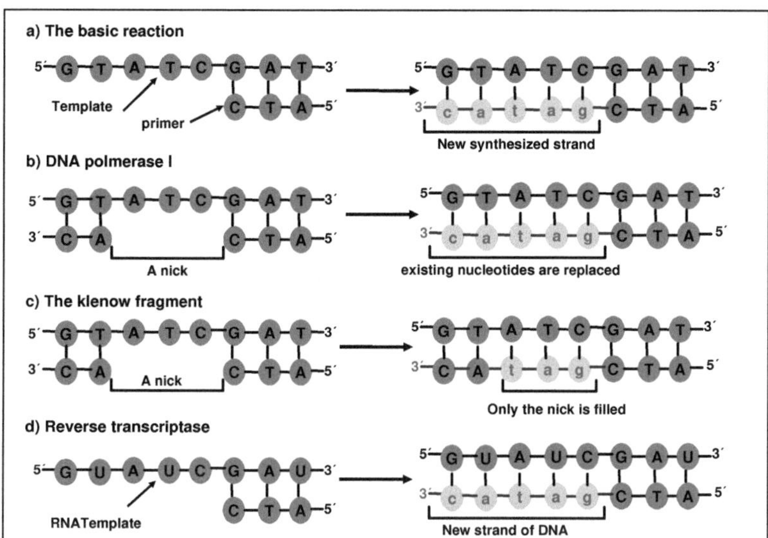

Fig. 2-9 The reactions catalyzed by DNA polymerases. a) **The basic reaction**: new DNA strand is synthesized in the 5`to 3` direction. b) **DNA polymerase I**, which initially fills in nicks and then continues to synthesize a new strand, degrading the existing one as it proceeds. c) **The klenow fragment**, which only fills in nicks. d) **reverse transcriptase**, which uses a RNA as a template.

3. Transformation (introduction of DNA into living cells)

Most species of bacteria are able to take up DNA molecules from the medium in which they grow. Often a DNA molecule taken up in this way will be degraded, but suddenly it is able to survive and replicate in the host cell. In particular this happens if the DNA molecule is a plasmid with an origin of replication recognized by the host.

Uptake and stable retention of a plasmid is usually detected by looking for expression of the genes carried by the plasmid. *E. coli* cells are normally sensitive to the growth inhibitory effects of the antibiotic's ampicillin and tetracycline. However, cells that contain the plasmid pBR322, one of the first cloning vectors to be developed back in the sets of genes, are resistant to these antibiotics. This is because pBR322 carries two sets of genes, one gene that codes for a β-lactamase enzyme that modifies ampicillin into a form that is non-toxic to the bacterium, and a second set of genes that code for enzymes that detoxify tetracycline. Uptake of pBR322 can be detected because of the *E. coli* cells are transformed from ampicillin and tetracycline sensitive ($amp^{S}tet^{S}$) to ampicillin and tetracycline resistant ($amp^{R}tet^{R}$).

In recent years the transformation has been extended to include uptake of any DNA molecule by any type of cell, regardless of whether the uptake results in a detectable change in the cell, or whether the cell, or whether the cell involved is bacterial, fungal, animal or plant. Most species of bacteria, including *E. coli*, take up only limited amounts of DNA under normal circumstances. In order to transform these species efficiently, the bacteria have to undergo some form of physical or chemical treatment that enhances their ability to take up DNA. Cells that have undergone this treatment are said to be competent.

3.1 Preparation of competent *E. coli* cells

It was observed that *E. coli* cells that had been soaked in an ice-cold salt solution were more efficient at DNA uptake than unsoaked cells. A treatment with calcium chloride ($CaCl_2$) is traditionally used, possibly causes the DNA to precipitate onto the outside of the cells, or perhaps the salt is responsible for some kind of change in the cell wall that improves DNA binding. In any case, soaking in $CaCl_2$ affects only DNA binding, and not the actual uptake into the cell. When DNA is added to treated cells, it remains attached to the cell exterior, and is not transported into the cytoplasm (Fig. 3-1). The actual movement of DNA into competent cells is stimulated by briefly raising the temperature 42 °C, the exact reason why this heat shock is effective is not understood.

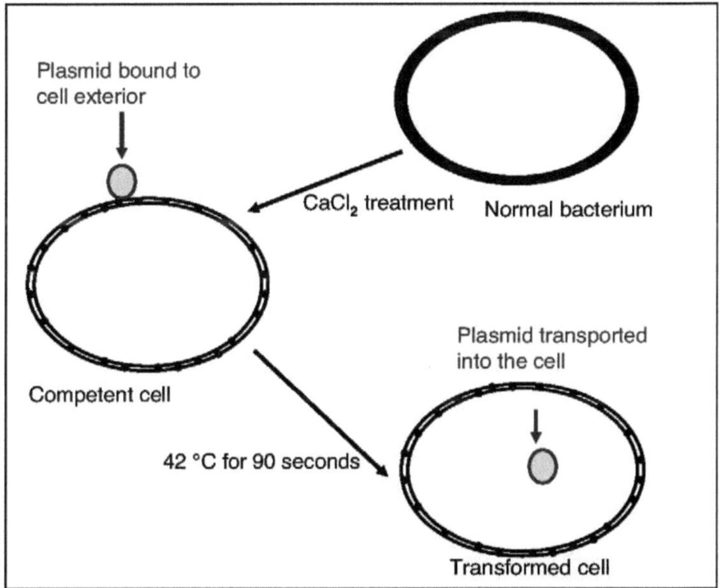

Fig. 3-1 The binding and uptake of DNA by a competent bacterial cell.

3.2 Selection of transformed cells

Normal *E. coli* cell (no plasmids) does not survive on agar containing antibiotic (ampicillin and tetracycline), but *E. coli* cell containing pBR322 plasmids survives and produces a colony on agar containing antibiotic (ampicillin and tetracycline). This is due to a selectable marker carried by the plasmid. A selectable marker is simply a gene that provides a transformed cell with a new characteristic, one that is not possessed by a non-transformant, as pBR322 has a selectable marker ampicillin resistance gene, only *E. coli* cells that have taken up a plasmid are $amp^R tet^R$ and able to form colonies on an agar medium that contains ampicillin or tetracycline; non-transformants whic are still $amp^s tet^s$, do not produce colonies on the selective medium.

Most plasmid cloning vectors carry at least one gene that confers antibiotic resistance on the host cells, with selection of transformants being achieved by plating onto an agar medium that contains the relevant antibiotic. The resistance gene on the plasmid must also be expressed, so that the enzyme that detoxifies tha antibiotic is synthesized. Expression of the resistance gene begins immediately after transformation, but it will be a few minutes before the cell the contains enough of the enzyme to be able to withstand the toxic effects of the antibiotic. For this reason, the transformed bacteria should not be plated onto selective medium immediately after the heat shock treatment, but first placed in a small volume of liquid medium, in the absence of antibiotic, and incubated for short time. Plasmid replication and expression can then get started, so that when the cells are plated out and encounter the antibiotic, they will already have synthesized sufficient resistance enzymes to be able to survive.

3.3 Cloning Vectors (Vehicles) for gene cloning: plasmids, bacteriophages and bacterial artificial chromosomes

The principles that govern the delivery of recombinant DNA in clonable form to a host cell, and its subsequent amplification in the host cell, are well illustrated by considering three popular cloning vectors commonly used in experiments with *E. coli*-plasmids, bacteriophages, and bacterial artificial chromosomes, and a vector used to clone large DNA segment in yeast.

"**Vector**" is an agent that can carry a DNA fragment into a host cell. If it is used for reproducing the DNA fragment, it is called a "**cloning vector**". If it is used for expressing certain gene in the DNA fragment, it is called an "**expression vector**" Commonly used vectors include **plasmid**, **Lambda phage**, **cosmid** and **yeast artificial chromosome (YAC).** Cloning vectors for E. coli, Yeast, fungi, plants and animals will be considered. A DNA molecule needs to display several features to be able to act as a vehicle for gene cloning, it must be able to replicate within the host cell, so that numerous copies of the recombinant DNA vehicle also needs to be relatively small, less than 10 kb in size, as large molecules tend to break down during purification, and are also more difficult to manipulate.

3.3.1 Cloning vectors based on *E. coli* plasmids

3.3.1.1 Basic features of plasmids

Plasmids are circular, double-stranded DNA molecules that replicate separately from the host chromosomes (independent existence in the bacterial cell, Fig. 1.14), due to **an *ori* sequence**, (origins of replication) which allows the plasmid to replicate in *E. coli* because it provides a DNA sequence that is recognized by the replication enzymes in the cells. Plasmids almost always carry one or more genes **(selectable marker or drug-resistance gene)**, and often these genes are responsible for a useful characteristic displayed by the host bacterium. For example, the ability to survive in normally toxic concentrations of antibiotics such as tetracycline (bind to ribosomal 30S subunit of the ribosome and inhibit protein synthesis), chloramphenicol (bind to the ribosomal 50S subunits and inhibit protein synthesis) or ampicillin-resistance gene (amp^r) (binds and inhibits a number of enzymes in the synthesis of cell wall as ß-galactosidase), due to the presence in the bacterium of a plasmid carrying antibiotic resistance genes. In the laboratory antibiotic resistance is often used as a selectable marker to ensure that bacteria in a culture contain a particular plasmid, and **unique restriction enzymes** cleavage sites. Sites present just once in the vector for the insertion of the DNA sequences that are to be cloned.

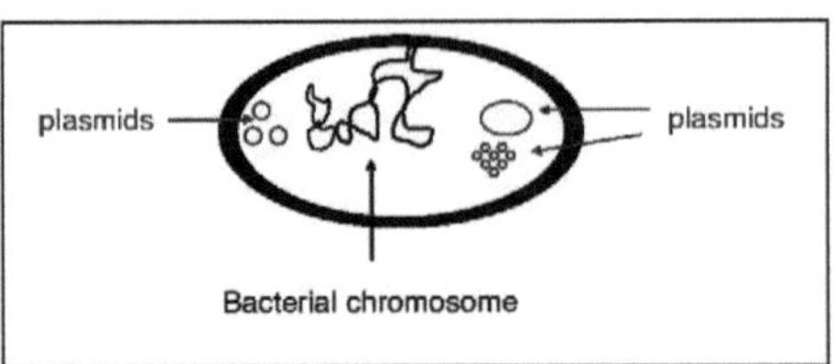

Fig. 3.2 Plasmids are independent genetic elements found in bacterial cells.

All plasmids possess at least one DNA sequence that can act as an origin of replication, so they are able to multiply within the cell quite independently of the main bacterial chromosome (Fig.3.3a). The smaller plasmids make use of the host cell's own DNA replicative enzymes in order to make copies of themselves, whereas some of the larger ones carry genes that code for special enzymes that are specific for plasmid replication.

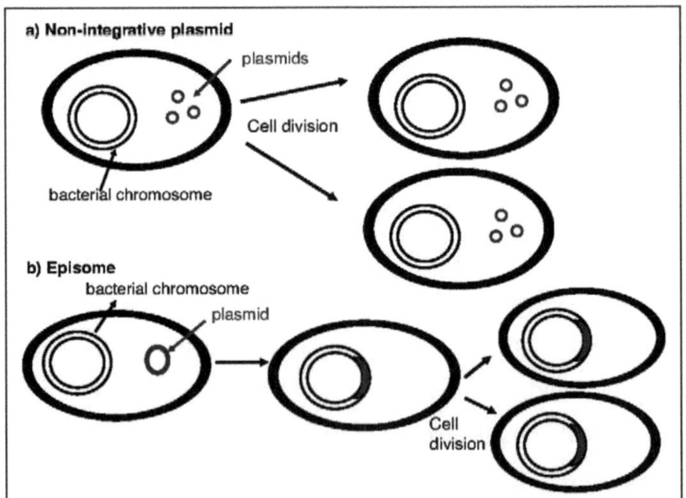

Fig. 3.3 Replication strategies a) a non-integrative plasmid, b) an episome.

A few types of plasmids are also able to replicate by inserting themselves into bacterial chromosome (Fig.3.3b). These integrative plasmids or episomes may be stably maintained in this form through numerous cell divisions, but will at some stage exist as independent elements. Integration is also an important feature of some bacteriophage chromosomes.

3.3.1.2 Size and copy number

The size and copy number of a plasmid are particularly important as far as cloning is concerned. Plasmids range from about 1.0 kb for the smallest to over 250 kb for the largest plasmids. A few are useful for cloning purposes. However, Large plasmids may be adapted for cloning under some circumstances. Naturally occurring bacterial plasmids range in size from 5,000 to 40,000 bp.

The copy number refers to the number of molecules of an individual plasmid that are normally found in a single bacterial cell. The factors that control copy number are not well understood, but each plasmid has a characteristic value that may be as low as one (especially for the large molecules, as Fig 3.4) ar as many as 50 or more to be present in the cell in multiple copies so that large quantities or the recombinant DNA molecule can be obtained (Fig. 3.5).

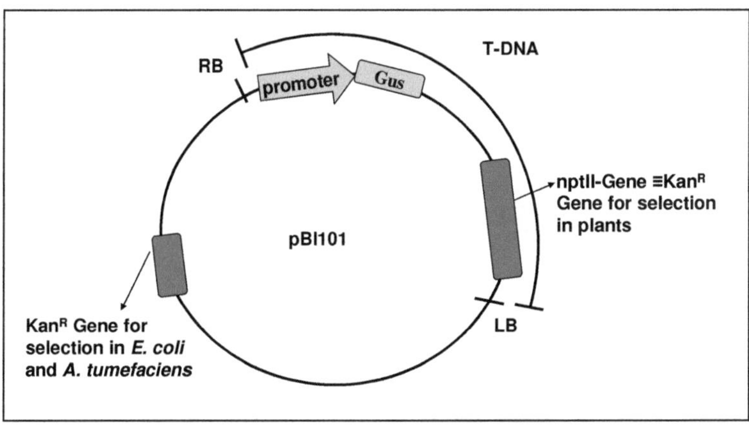

Fig. 3. 4 Low copy plasmid ~ 12.5 kb and restriction Map of pBI101 (pBIB-Kan^R).

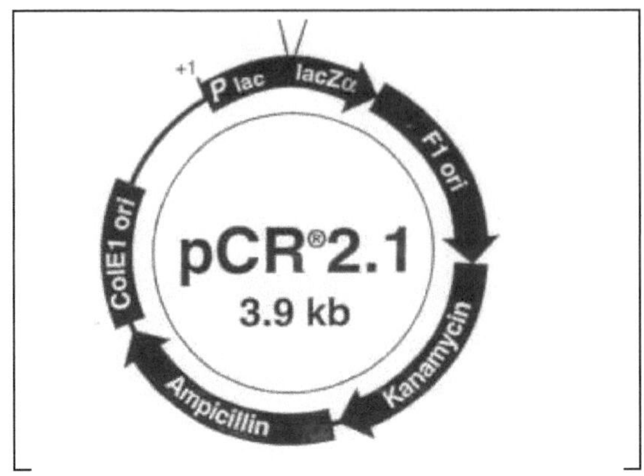

Fig. 3.5 High copy plasmid ~3.9 kb and restriction Map of pCR2.1.

A plasmid vector is made from natural plasmids by removing unnecessary segments and adding essential sequences. To clone a DNA sample, the same restriction enzyme must be used to cut both the vector and the DNA sample. Therefore, a vector usually contains a sequence (**polylinker**) which can recognize several restriction enzymes so that the vector can be used for cloning a varicty of DNA samples as in Fig. 3.6.

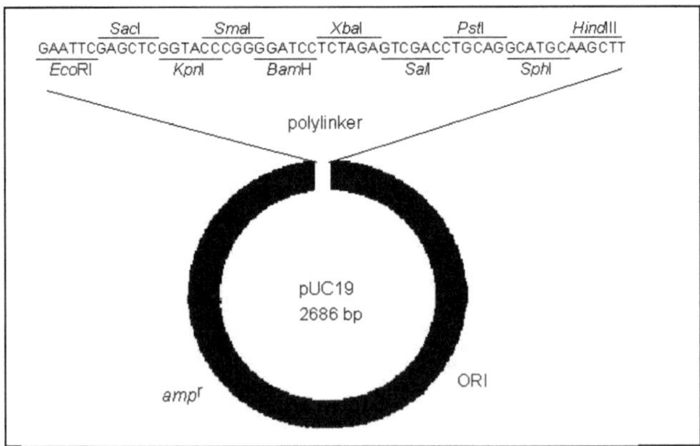

Fig. 3.6 A typical plasmid vector. It contains a polylinker which can recognize several different restriction enzymes, an ampicillin-resistance gene (amp^r) for selective amplify-ication, and a replication origin (ORI) for proliferation in the host cell.

3.3.1.3 Conjugation and compatibility

Plasmids fall into two groups: conjugative and non-conjugative. Conjugative plasmids are characterized by the ability to promote sexual conjugation between bacterial cells. A process that can result in a conjugative plasmid spreading from one cell to all the other cells in a bacterial culture. Conjugation and plasmid transfer are controlled by a set of transfer or tra genes, which are present on conjugative plasmids but absent from the non-conjugative type. However, a non-conjugative plasmid may, under some circumstances, be cotransferred along with a conjugative plasmid when both are present in the same cell. Several different kinds of plasmids may be found in a single cell, including more than one different conjugative plasmid at any one time. In fact, cells of *E. coli* have been known to contain up to seven different plasmids at once. To be able to coexist in the same cell, different plasmids must be compatible. If two plasmids are incompatible then one or other will be quite rapidly lost from the cell. Different types of plasmid can be a ssigned to different incompatibility groups on the basis of whether or not they can coexist, and plasmids from a single incompatibility group are often related to each other in various ways.

3.3.1.4 Plasmid classification

The classification of plasmids is based on the main characteristic coded by the plasmid genes. The five main types of plasmid:

1- **Fertility or 'F' plasmids** carry only tra genes and have no characteristic beyond the ability to promote conjugal transfer of plasmids, e.g. F plasmid of *E. coli*.

2- **Resistance or 'R' plasmids** carry genes conferring on the host bacterium resistance to one or more antibacterial agents, such as chloramphenicol, ampicillin and mercurry. R plasmids are very important in clinical microbiology.

3- **Col plasmids** code for colicins, proteins that kill other bacteria, e.g. ColE1 or *E. coli*.

4- **Degradative plasmids** allow the host bacterium to metabolize unusual molecules such as toluene and salicylic acid, e.g. TOL of *Pseudomonas putida*.

5- **Virulence plasmids** confer pathogenicity on the host bacterium; e.g. **Ti plasmids** of ***Agrobacterium tumefaciens***, which induce crown gall disease on dicotyledonous plants.

3.3.2 Cloning vectors based on M13 bacteriophage

The most essential requirement for any cloning vector is that it has a means of replicating in the host cell. Bacteriophages, or phages as they are commonly known, are viruses that specifically infect bacteria, like all viruses, phages are very simple in structure, consisting merely of a DNA (or occasionally RNA) molecule carrying a number of genes, including several for replication of the phage, surrounded by a protective coat or capsid made up of protein molecules. With bacteriophages such as M13 the situation as regards replication more complex. Phage DNA molecules generally carry several genes that are essential for replication, including genes coding for components of the phage protein coat and phage-specific DNA replicative enzymes.

3.3.3 Cloning vectors based on Lambda (λ) bacteriophage

λ phages are viruses that can infect bacteria. The major advantage of the λ phage vector is its high transformation efficiency, about 1000 times more efficient than the plasmid vector. The extreme ends of the λ DNA are known as **COS sites**, each is single stranded, 12 nucleotides long. Because their sequences are complementary to each other, one end of λ DNA may base-pair with the other end of a different λ DNA, forming concatemers. The two ends of a λ DNA may also bind together, forming a circular DNA. In the host cell, the λ DNA circularizes because ligase may seal the join of the COS sites. In the assembly process of λ virions, two proteins Nu1 and A can recognize the COS site, directing the insertion of the λ DNA between them into an empty head.

The filled head is then attached to the tail, forming a complete λ virion. The whole process normally takes place in the host cell. However, to prepare the λ virion carrying recombinant λ DNA, the following ***in vitro* assembly system** is commonly used. Proteins Nu1 and A are encoded by the genes in the λ genome. If the two genes are mutated, λ DNA cannot be packaged into the pre-assembled head. Because tails attach only to filled heads, the cell will accumulate separate empty heads and tails, which can then be extracted. When the extract is mixed with recombinant λ DNA and proteins Nu1 and A, the complete λ virion carrying recombinant λ DNA will be assembled as in Fig. 3.7.

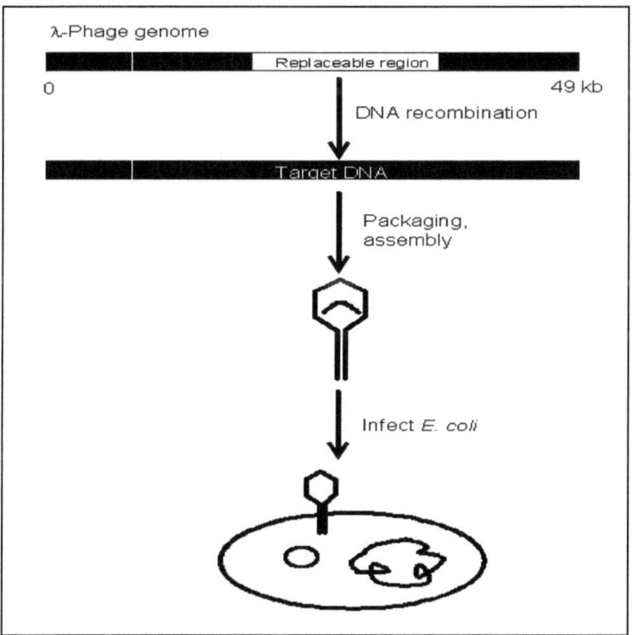

Fig. 3.7 Schematic drawing of the DNA cloning using λ phages as vectors. The DNA to be cloned is first inserted into the λ DNA, replacing a nonessential region. Then, by an in vitro assembly system (described below), the λ virion carrying the recombinant DNA can be formed. The λ genome is 49 kb in length which can carry up to 25 kb foreign DNA.

3.3.4 Cloning vector based on Cosmid

The cosmid vector is a combination of the plasmid vector and the COS site which allows the target DNA to be inserted into the λ head. It has the following advantages:

- High transformation efficiency.
- The cosmid vector can carry up to 45 kb whereas plasmid and λ phage vectors are limited to 25 kb.
- Some genes from higher eukaryotes may be as long as 30 kb, can be cloned in one piece using cosmids.

The cos site is the site at which multiple copies of the genome, attached in one long piece called a concatamer.

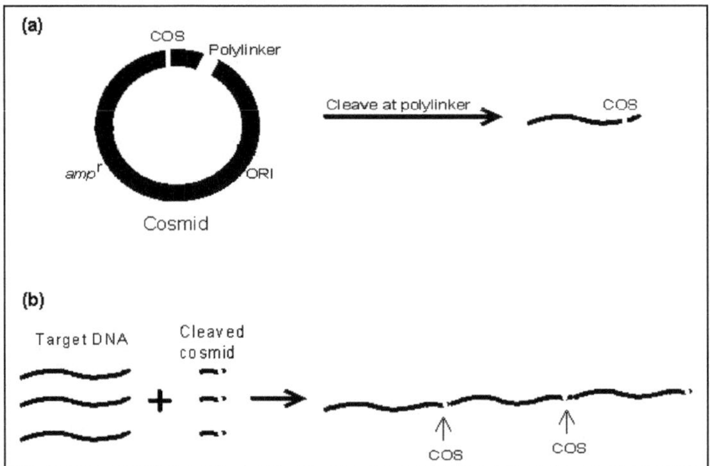

Fig. 3.8 Cloning by using cosmid vectors. (a) In addition to *amp*^{r}, ORI, and polylinker as in the plasmid vector, the cosmid vector also contains a COS site. **(b)** After cosmid vectors are cleaved with restriction enzyme, they are ligated with DNA fragments. The subsequent assembly and transformation steps are the same as cloning with λ phages.

3.3.5 Cloning vectors based on YAC (yeast artificial chromosome)

The yeast artificial chromosome (YAC) vector is capable of carrying a large DNA fragment (up to 2 Mb), but its transformation efficiency is very low

3.3.5.1 Essential components of YAC vectors

- Centromers (CEN), telomeres (TEL) and autonomous replicating sequence (ARS) for proliferation in the host cell.
- *amp*^{r} for selective amplification and markers such as TRP1 and URA3 for identifying cells containing the YAC vector.
- Recognition sites of restriction enzymes (e.g., *EcoRI* and *BamHI*).

Shuttle vector A vector that can replicate in the cells of more than one organism (e.g. *E. coli* and in Yeast).

4. DNA Sequence

4.1 Restriction digest DNA in the laboratory

The required amount of DNA must be pipetted into a test tube. The amount of DNA that will be restricted depends on the nature of the experiment and the concentration of DNA. Before adding the restriction enzyme, must be added the suitable buffer. Type II restriction endonucleases require Mg^{2+} in order to function, and NaCl for ionic strength. It is also advisable to add a reducing agent, such as dithiothreitol (DTT), which stabilizes the enzyme and prevents its inactivation. Providing the right conditions for the enzyme is very important and incorrect NaCl or Mg^{2+} concentrations may not only decrease the activity of the restriction endonuclease, but may also cause changes in the specificity of the enzyme, so that DNA cleavage occurs at additional, non-standard recognition sequences. The last factor to consider is the incubation temperature. Most restriction endonucleases work the best at 37°C, but a few have different requirements. *Taq* DNA polymerase, has a high working temperature. Restriction digests with *Taq* must be incubated at 72°C to obtain maximum enzyme activity, but other *SmaI* enzyme, has a low working temperature. Restriction digests with *SmaI* must be incubated at 28°C to obtain maximum enzyme activity. After 1-2 hour the restriction should be completed, and the DNA fragments produced by restriction are to be used in cloning experiments, the enzyme must somehow be destroyed so that it does not accidentally digest other DNA molecules that may be added at a later stage. They are several ways of killing the enzyme. For many a short incubation (2-3 minutes) at 65°C is sufficient, for others phenol extraction or the addition of ethylenediaminetatraacetic acid (EDTA), which binds Mg^{2+} ions preventing restriction endonuclease action is used. A restriction digest results in a number of DNA fragments, the sizes of which depend on the exact positions of the recognition sequences for the endonuclease in the original molecule.

DNA fragments produced by restriction endonuclease cleavage. The next step in restriction analysis to a construct a map showing the relative positions on the DNA molecule of the recognition sequences for a number of different enzymes. Only when a restriction map is available can the correct restriction endonucleases be selected for the particular cutting manipulation that is required. To construct a restriction map, a series of restriction digests must be performed. First, the number and the sizes of the fragments produced by each restriction endonuclease must be determined by gel electrophoresis, followed by comparison with size markers. This information must then supplement by a series of single digestion (cut by one restriction endonuclease) or double digestions, in which the DNA is cut by two restriction endonucleases at once. It may be possible to perform a double digestion in one step, if both enzymes have similar requirements for pH, Mg^{2+}, concentration. Alternatively, the two digestions may have to be carried out after the other, adjusting the reaction mixture after the first digest to provide a different set of conditions for the second enzyme. Comparing the results of single and double digests will many, if not all, of the restriction sites to be mapped. These can usually be resolved by partial digestion, carried out under conditions that result in cleavage of only a limited number of the restriction sites on any DNA molecule. Partial digestion is usually achieved by reducing the incubation period, so the enzyme does not have time to cut all the restriction sites, or by incubating at a low temperature (e.g. 4 °C rather than 37 °C) or a high temperature for short time, which limits the activity of the enzyme. However, the sizes indicate which restriction fragments in the complete digest are next to one another in the uncut molecule.

4.2 Separation of molecules by gel electrophoresis

The gel electrophoresis was developed at 1970s. DNA molecules, like Proteins and many other biological compounds, carry an electric charge, negative in the case of DNA. Consequently, when DNA molecules are placed in an electric field they migrate towards the positive pole. The rate of migration of a molecule depends on two factors, its shape and its charge-to-mass ratio. Unfortunately, most DNA molecules are the same shape and all have very similar charge to mass ratios. The size of the DNA molecule is a factor, when the electrophoresis is performed in a gel. A gel, which is usually made of agarose or polyacrylamide or a mixture of the two, comprises a complex network of pores, through which DNA molecules must travel to reach the positive electrode. Polyacrymalide gels are used to separate fragments containing up to about 1000 base pairs, whereas more porous agarose gels are used to resolve mixtures of larger fragments (up to about 20 kb). An important feature of these gels is their high resolving power. The smaller the DNA molecule, the faster it can migrate the gel. Gel electrophoresis therefore separates DNA molecules according to their size (Fig. 4.1).

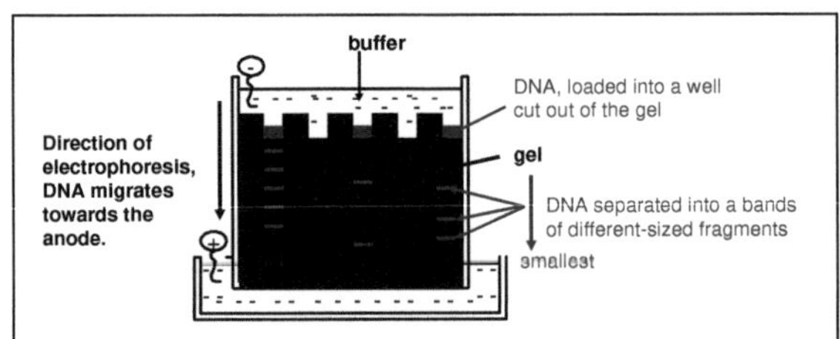

Fig. 4.1 Apparatus for slab-gel electrophoresis capable of running samples. Showing the bands under ultraviolet after staining with ethidium bromide in the presence of standard marker to identify the size of bands.

In practice the composition of the gel determines the sizes of the DNA molecules that can be separated. A 0.5 cm thick slab of 0.5 % agarose, which has relatively large pores, would be used for molecules in the size range 5-15 kb.

4.3 Visualizing DNA molecules in a gel by staining

Ethidium bromide (EtBr) [see Fig. 4.2] can be used to separate supercoiled DNA from non-supercoiled molecules. Ethidium bromide binds to DNA molecules by intercalating between adjacent base pairs, causing partial unwinding of the double helix, this unwinding results in a decrease in the buoyant density. However supercoiled DNA, with no free ends, has very little freedom to unwind, and can only bind a limited amount of EtBr. Bands showing the position of the different size classes of DNA fragment are clearly visible (an intense orange) under ultraviolet irridiation after EtBr staining. This procedure is very hazardous because ethidium bromide is a powerful mutagen and the ultraviolet radiation used to visualize the DNA can cause severe burns. A band containing only 50-10 ng of DNA can readily be seen (Fig. 4.1).

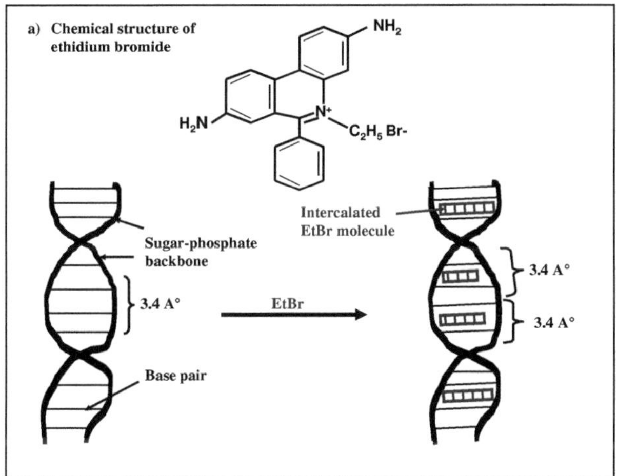

Fig. 4.2 Partial unwinding of the DNA double helix by EtBr intercalation between adjacent base pairs.

4.3.1 Autoradiography of radioactively labelled DNA

The staining is a limit to its sensitivity. If the amount DNA less than about 10 ng of DNA is present per band, it is unlikely that the bands will show up after staining. For small amounts a more sensitive detection method is needed as autoradiography. If the DNA is labelled before electrophoresis, by incorporation of a radioactive marker into the individual molecules, the DNA can be visualized by placing an X-ray sensitive photographic film over the gel. The radioactive DNA exposes the film, revealing the banding pattern. A DNA molecule is usually labelled by incorporating nucleotides that carry a radioactive isotope of phosphorus ^{32}P (Fig. 4.3). Several methods are available, the most popular being nick translation and end filling.

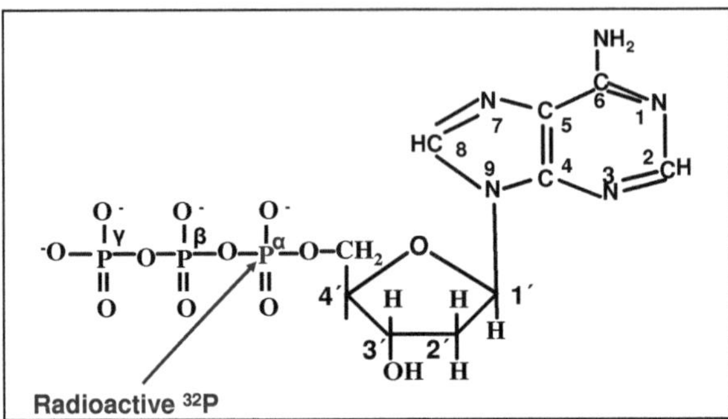

Fig. 4.3 Structure of radioactive labelled of α-32P-deoxyadenosine triphosphate (α- 32P- dATP).

Nick translation refers to the activity of DNA polymerase I. Most purified samples of DNA contain some nicked molecules, however carefully the preparation has been carried out, which means that DNA polymerase I is able to attach to the DNA and catalyse a strand replacement reaction. This reaction requires a supply of nucleotides: if one of these is radioactively labelled, the DNA molecule will itself become labelled. Nick translation can be used to label any DNA molecule but may under some circumstances also cause DNA cleavage. End filling is a gentler method that rarely causes breakage of the DNA, but unfortunately can only be used to label DNA molecules that have sticky ends. The enzyme used is the klenow fragment, which fills in a sticky end by synthesizing the complementary strand, and the filling reaction is carried out in the presence of labelled nucleotides, the DNA itself becomes labelled. Very small amount label DNA can be detected in gels by autoradiography. As little as 2 ng of DNA per band can be visualized under ideal conditions.

Two different techniques were developed almost simultaneously; the chain termination method by F. Sanger and A.R. Coulson (1977) in the UK, and the chemical degradation method by A. Maxam and W. Gilbert in the USA. The two techniques are radically different but equally valuable. Both allow DNA sequences of several kilobases in length to be determined in the minimum of time.

4.3.2 Sanger dideoxy sequencing method (the chain termination)

The chain termination method requires single-stranded DNA and so the molecule to be sequenced is usually clone into vector. This is because chain termination sequencing involves the enzymatic synthesis of a second strand of DNA, complementary to an existing template. The first step in a chain termination sequencing experiment is to anneal a short oligonucleotide primer, these primer acts as the starting point for the complementary strand synthesis reaction carried out by the klenow fragment of DNA polymerase I, these enzymes need a double-stranded region from which to begin strand synthesis. The complementary strand synthesis reaction is started by adding the enzyme plus each of the four deoxynucleotides [2`-deoxyadenosine 5`-triphosphate (dATP), 2`-deoxyguanosine 5`-triphosphate (dGTP), 2`-deoxythymidine 5`-triphosphate (dTTP), 2`-deoxycytidine 5`-triphosphate (dCTP)]. In addition, a single modified nucleotide is also included in the reaction mixture. This is a 2`, 3`dideoxynucleotide analog (e.g. dideoxy ATP) which can incorporate into the growing polynucleotide strand just as efficiently as the normal nucleotide, but which blocks further strand synthesis. This is because the dideoxynucleotide lacks the hydroxyl group at the 3`position of the sugar component needed to form the next phosphodiester bond (Fig. 4.4). This group is needed for the next nucleotide to be attached; chain termination therefore occurs whenever a dideoxynucleotide is incorporated by the enzyme.

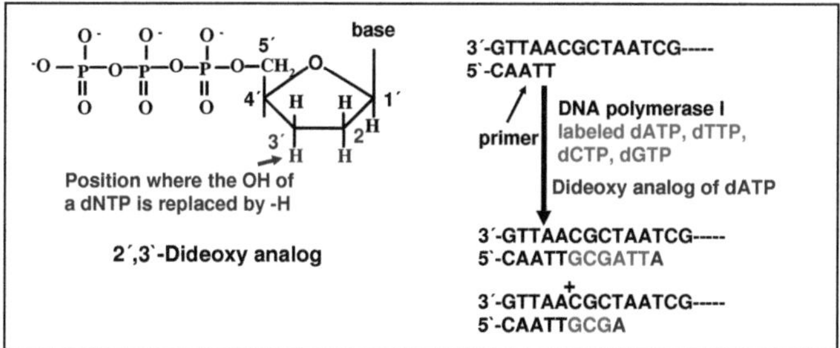

Fig. 4.4 chain –termination DNA sequencing. Fragments are produced by adding the 2`, 3`-dideoxy analog of a dNTP to each of four polymerization mixtures. For example, the addition of the dideoxy analog of dATP results in fragments ending in A.

If dideoxy ATP is added to the reaction mix termination occurs at positions opposite thymidines in the template (Fig. 4.4). But the termination does not always occur at the first T as normal dATP is also present and may be incorporated instead of the dideoxynucleotide. The ration of dATP to dideoxyATP is such that an individual strand can be polymerized for a considerable distance before a dideoxyATP molecule is added. The result is that a family of new strands is obtained, all of different lengths, but each ending in dideoxyATP. The strand synthesis reaction is carried out four times in parallel. As well as the reaction with dideoxyATP, there is one dideoxyTTP, one with dideoxyGTP, and one with dideoxyCTP. The result is four distinct families of newly synthesized polynucleotides, one family containing strands all ending in dideoxyATP, one of strands ending in dideoxyTTP, etc.

The next step is to separate the components of each family so the lengths of each strand can be determined. This can be achieved by gel electrophoresis, the gels contain urea, which denatures the DNA so the newly synthesized strands dissociate from the templates. In addition, the electrophoresis is carried out at a high voltage, so the gel heats up to 60°C and above, making sure the strands do not reassociate in any way. Each band in the gel contains only a small amount of DNA, so autoradiography has to be used to visualized the resuls. The label is introduced into the new strands by including a radioactive deoxynucleotide (e.g. ^{32}P). Reading the sequence is very easy. First the band that has moved the furthest is located. This represents the smallest piece of DNA, the strand terminated by incorporation of the dideoxynucleotide at the first position in the template (Fig. 4.5).

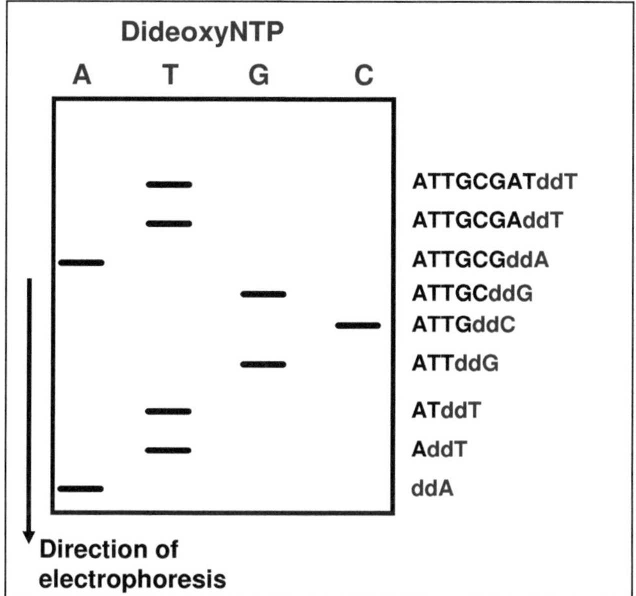

Fig. 4.5 Interpreting the autoradiograph produced by chain termination sequencing experiment. Each track contains the fragments produced by strand synthesis in the presence of one of four Dideoxy nucleotide triphosphates (dideoxyNTPs). The sequence is read by identifying the track that each fragment lies in starting with the one that has moved the furthest, and gradually progressing up through the autordiograph.

The process is continued along the autoradiograph until the individual bands become so bunched up that they cannot be separated from one another. Generally, it is possible to read a sequence of about 400 nucleotides from one autoradiography.

Fluorescence detection is a highly effective alternative to autoradiography. A fluorescenct tag is attached to an oligonucleotide primer-a differently coloured one in each of the four chain-terminating reaction mixtures (e.g., a blue emitter for termination at A and a red one of termination at C). The reaction mixtures are combined and electrophoresed together. The separated bands of DNA are then detected by their fluorescence as they emerge from the gel; the sequence of their colours directly gives the base sequence. Sequences of up to 500 bases can be determined in this way. Fluorescence detection is attractive because it eliminates the use of radioactive reagents and can readily be automated.

4.3.3 Maxam and Gilbert sequencing Technique (Chemical cleavage)

There are only a few similarities between the Sanger-Coulson and Maxam-Gilbert methods of DNA sequencing. The Maxam-Gilbert method requires double stranded DNA fragments, so cloning into vector is not an essential first step. Neither is a primer needed, because the basis of the Maxam-Gilbert technique is not synthesis of a new strand, but cleavage of the existing DNA molecule using chemical reagents that act specifically at a particular nucleotide. There are several variations of the Maxam-Gilbert method, differing in details such as the way in which the labelled DNA is obtained, and the exact nature of the cleavage reagents that are used. Most of these reagents are very toxic-chemicals that cleave DNA molecules in the test tube. Allan Maxam and Walter Gilbert starts with a single-stranded DNA that is labeled at one end with ^{32}P. Polynucleotide kinase is usually used to add ^{32}P at the 5`-hydroxyl terminus. The labeled DNA is then broken preferentially at one of the four nucleotides. The conditions are chosen so that an average of one break is made per chain. In the reaction mixture for a given base, then, each broken chain yields a radioactive fragment extending from the ^{32}P label to one of the positions of that base, and such fragments are produced for every position of the base. For example, if the sequence is

5`-^{32}P-GCTACGTA-3´

The radioactive fragments produced by specific cleavage on the 5`side of each of the four bases would be

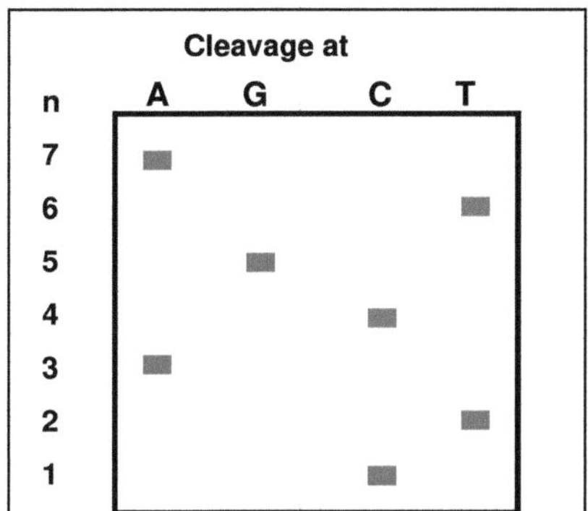

Cleavage at A: ^{32}P-GCT
^{32}P-GCTACGT

Cleavage at G: ^{32}P-GCTAC

Cleavage at C: ^{32}P-G
^{32}P-GCTA

Cleavage at T: ^{32}P-GC
^{32}P-GCTACG

Fig. 4.6 Schematic digram of a gel showing the radioactive fragments formed. By specific cleavage of 5`-^{32}P-GCTACGTA-3´ at each of the four bases (A, G, C, and T lanes). The number of nucleotides (n) in a fragment is shown on the left. The base sequence is read in ascending order.

The fragment in each mixture are then separated by polyacrylamide gel electrophoresis, which can resolve DNA molecules differing in length by just one nucleotide. The next step is to look at an autoradiogram of this gel. In this idealized example (Fig. 4.6), the lowest band would be in the C lane, and the next one up in the T lane, followed by one in the A lane. Hence, the sequence of the first three nucleotides detected is 5`-CTA-3`. Reading all seven bands in ascending order gives the sequence 5`-CTACGTA-3`. Thus, the autoradiogram of a gel produced from four different chemical cleavages displays a pattern of bands from which the sequence can be read directly.

In practice, DNA is specifically cleaved by reagents that modify and then remove certain bases from their sugars. Purines are damaged by dimethylsulfate, which methylates guanine at N-7 and adenine at N-3. The glycosidic bond of a methylated purine is readily broken by heating or neutral pH, which leaves the sugar without a base. Subsequent heating in alkali leads to cleavage of the backbone at G, whereas treatment with diluted acid causes cleavage at both A and G. Cytosine and thymine are split off by hydrazine. The backbone is then cleaved at both C and T by piperidine. The bands visualized by an autoradiogrphy represent these labelled fragments, and the nucleotide sequence can now be read from the autoradiograph.

4.4 Sequencing PCR products

It is necessary to determine the sequence of the amplified DNA fragment. This can be achieved by cloning the PCR product and using a standard chain termination or chemical degradation sequencing method as described above.

One problem with this approach is that the sequences of individual clones might not faithfully represent the sequence of the original template DNA molecule, because of the errors that are occasionally introduced by Taq polymerase during the amplification process. Methods that sequence a PCR product directly, without the need for cloning, are therefore required.

4.4.1 Direct sequencing of PCR products

The direct sequencing method is based on the Sanger-Coulson technique and therefore requires single-stranded DNA as the starting material. The PCR product is of course double-stranded, so some means of purifying single strands is needed. The initial PCR with one normal and one modified primer, the modified primer altered in such a way that the DNA strands synthesized from it are easily purified. A clever way of doing this is by attaching small magentic beads to one of the primers. After the PCR, single-stranded DNA is obtained by separating the magnetic strand from the ordinary strand. Or may be used a biotin-labelled primer, with the single strands separated by binding to avidin, a protein that has a high affinity for biotin. Once single-stranded DNA has been purified, the remainder of the procedure is similar to the standard Sanger-Coulson method, in which families of chain-terminated molecules are synthesized by a DNA polymerase such as Sequenase. But in PCR product, the primer has to be complementary to the purified single strands, and therefore is the primer that was not labelled with magnetic beads or with biotin.

4.4.2 Thermal cycle sequencing

It is always necessary to purify single strands in order to sequence a PCR product. This is no, because it is possible to combine PCR and DNA sequencing into a single reaction, resulting in the technique called thermal cycle sequencing. In thermal cycle sequencing the reaction mixture that is prepared in similar to a PCR mixture but with two important exceptions. The first is that only one primer is added to the reaction, which means that the amplification process characteristic of PCR can not occur, and instead all that happens is that one strand of the template DNA is copied. However, this copying is repeated many times because the reaction is thermally cycled, exactly as in a real PCR, and the enzyme that does the copying is a thermostable one such as *Taq* polymerase.

The second difference between a thermal cycle sequencing reaction and a PCR is that the former is carried out four times in parallel, each time with a different dideoxynucleotide included in the reaction. The new molecules that are synthesized are therefore chain-terminated, and the sequence can be read as for a standard Sanger-Coulson experiment.

4.5 Computer aided sequence analysis

Computers are now routinely used to store DNA sequences. Search programs were used for sequence comparisons on DNA and amino acid sequences obtained from GenBank (http://www.ncbi.nlm.nih.gov/), EMBL (http://www.ebi.ac.uk/embl), and SWISS-PROT (http://www.ebi.ac.uk/ swissprot /index.html) databases. Computational analysis for DNA sequence data was performed with the Wisconsins Genetics Computer Group Software package GCG for restriction ezyme recognition sites, start and stop signals. The FASTA and Basic Local Alignment Search Tool (BLAST). Sequence alignments were performed using the BESTFIT program of the GCG package, alignX (Vector NTI, infor Max, Version 7), and alignment of amino acid sequences was established by using the BCM Search Launcher.

Computational biology involves the use of techniques including applied mathematics, informatics, statistics, computer science, artificial intelligence, chemistry, and biochemistry to solve biological problems usually on the molecular level.

1. The use of computer systems to aid genomics and post-genomics research;

2. Computerized assembly of sequence contigs, examination of sequences for the presence of

genes, prediction of gene function, and storage of the vast amounts of data that are generated during a genome project.

4.6. Edmann degradation for protein

The amino acid sequence of a protein is determined by identifying amino acid residues as they are sequentially cleaved from the intact protein. Sequence analysis of nucleic acids is based on the generation of sets of DNA or RNA fragments with common ends and the separation of these oligonucleotide fragments by polyacrylamide electrophoresis.

Let us consider how the sequence of a short peptide, such as:.

Ala-Gly-Asp-Phe-Arg-Gly

Could be established. First, the amino acid composition of the peptide is determined. The peptide is hydrolyzed into its constituent amino acids by heating it in 6N HCl at 110°C for 24 hours. Stanford Moore and William Stein showed that amino acids in hydrolysates can be separated by ion-exchange chromatography on columns of sulfonated polystyrene and quantitated by reacting them with ninhydrin. Amino acids treated this way give an intense blue color, except for proline, which gives a yellow color because it contains a secondary amino group. The concentration of amino acid in a solution is proportional to the optical absorbance of the solution after heating it with ninhydrin. This technique can detect a microgram (10 nmol) of an amino acid, which is about the amount present in a thumbprint. As little as a nanogram (10 pmol) of an amino acid can be detected by means of fluorescamine, which reacts with the α-amino group to form a highly fluorescent product.

The amino terminal residue of a protein or peptide can be identified by labeling it with a compound that forms a stable covalent link. Fluorodinitrobenzene (FDNB) was first used for this purpose by Frederick Sanger. Dabsyl chloride is now commonly used because it forms intensely colored derivatives that can be detected with high sensitivity. It reacts with an uncharged α-NH_2 group to form sulfonamide derivative that is stable under conditions that hydrolyze peptide bonds. Although the dabsyl method for determining the amino-terminal residue is sensitive and powerful, it can not be used repeatedly on the same peptide because the peptide is totally degraded in the caid hydrolysis step. Pehr Edman devised a method for labeling the amino-terminal residue and cleaving it from the peptide without disrupting the peptide bonds between the other amino acid residues. The Edman degradation sequentially removes one residue at a time from the amino end of a peptide. Phenyl isothiocyanate reacts with the uncharged terminal amino group of the peptide to form a phenylthiocarbamoyl derivative. Then, under mildly acidic conditions, a cyclic derivative of the terminal amino acid is liberated, which leaves an intact peptide shortened by one amino acid. The cyclic compound is a phenylthiohydantoin (PTH)-amino acid, which can be identified by chromatographic procedures. Furthermore, the amino acid composition of the shortened peptide.

Analyses of protein structures have been markedly accelerated by the development of sequenators, which are automated instruments for the determination of amino acid sequence. In a liquid-phase sequenator, a thin film of protein in a spinning cylindrical cup is subjected to the Edman degradation. The reagents and extracting solvents are passed over the immobilized film of protein, and the released PTH-amino acid is identified by high-pressure liquid chromatography (also called high-performance liquid chromatography, HPLC). One cycle of the Edman degradation is carried out in less than two hours. By repeated degradations, the amino acid sequence of some fifty residues in a protein can be determined. Gas-phase sequenators can analyze picomole quantities of peptides and proteins. This high sensitivity makes it feasible to analyze the sequence of a protein sample eluted from a single band of an SDS-polyacrylamide gel (Fig. 4.7).

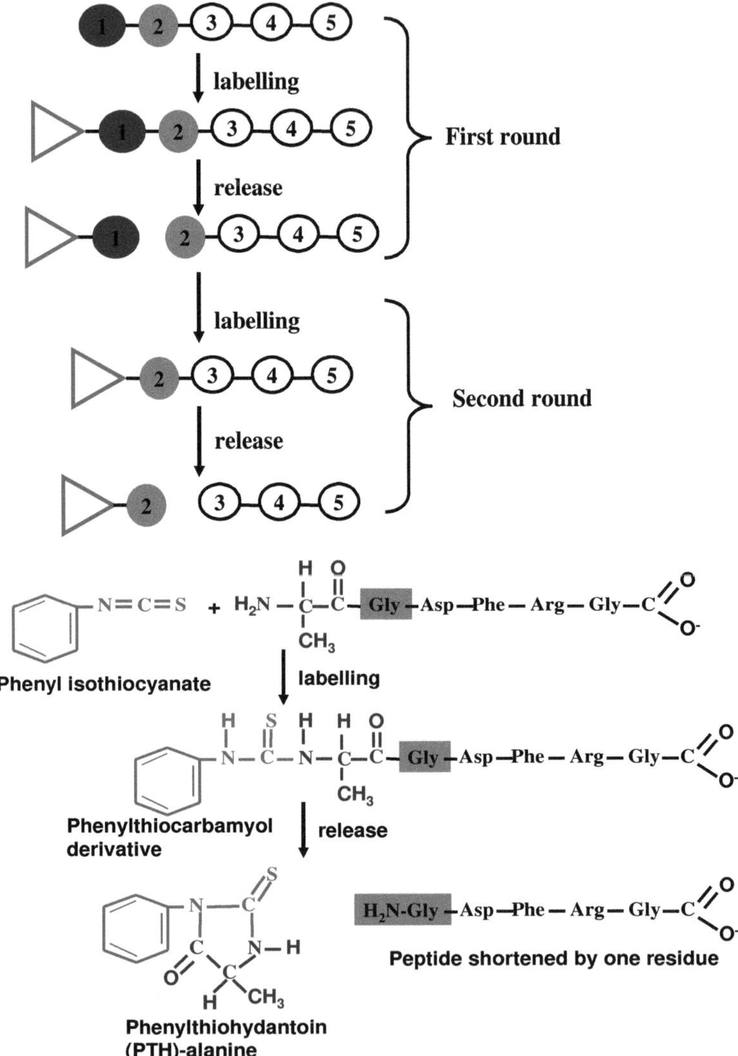

Fig. 4-7: The Edman degradation. The labeled amino-terminal residue (PTH-alanine in the first round) can be released without hydrolyzing the rest of the peptide. Hence, the amino terminal residue of the shortened peptide (Gly-Asp-Phe-Arg-Gly) can be determined in the second round. Three more rounds of the Edman degradation reveal the complete sequence of the original peptide.

5. What is bioinformatics?

Bioinformatics is the application of computer technology to the management of biological information. Computers are used to gather, store, analyze and integrate biological and genetic information which can then be applied to gene-based drug discovery and development. The science of bioniformatics (algorithms on sequence), which is the melding of molecular biology with computer science, is essential to the use of genomic information in understanding human diseases and in the identification of new molecular targets for drug discovery. In recognition of this, many universities, pharmaceutical firms and government institutions have formed bioinformatics groups, consisting of computational biologists and bioinformatics computer scientists to solve the mass of information generated by large scale sequencing efforts underway in laboratories around the world.

Bioinformatics and computational biology involve the use of techniques including applied mathematics, informatics, statistics, computer science, artificial intelligence, chemistry, and biochemistry to solve biological problems usually on the molecular level.

5.1 Bioinformatics at the heart of a scientific revolution

Parallel progress of several technologies allying minaturisation and robotization: Data produced at an industrial rhythm (sequencing for genomics, biochips and 2D gels for post-genomics)

- "Giga" Data Management
  - memory Go;
  - clock frequency GHZ
  - networks Gb/s
- Usable results
  - combinatorial chemistry;
  - genetic therapies
- Understanding of biological systems at a molecular level (end of reduction).

5.2 Bioinformatics are three things:

1. Standard service in computer science (biocomputing) as Data management: data bases, standard statistical data analysis, image analysis.

2. Specific data analysis (computational biology) as modelling: to assist the understanding of biological mechanisms from molecular data (sequences, structures and expression level of macromolecules).

3. Computing with the help of biomolecules (biomolecular computing, DNA computing), intensive parallel computing using DNA, (high storage capacity, low energy consumption).

5.3 To understand a genome, three distinct types of analysis must be carried out

a. What are the needs for genomics?

1. Manage all this information;
2. Integrate, add some semantics: from information to knowledge;
3. Raw data are sequences: importance of algorithmics on words.

b. What are the needs for post-genomics or functional genomics?

1. Manage all this information;
2. Integrate (interaction networks, "Systenomics").
3. Raw data are numbers: importance of data analysis (normalization, clustering).

c. What are the needs for Bioinformatics?

1. The use of computer systems to aid genomics and post-genomics research;
2. Computerized assembly of sequence contigs, examination of sequences for the presence of genes, prediction of gene function, and storage of the vast amounts of data that are generated during a genome project.

5.4 What is Bioinformatics used for?

Bioinformatics is used for a virtally limitless number of tasks, but some of the most common are:

1- Finding homologs (´twins´) of a gene in your favourite species given a sequence you have in a model species. eg. Finding a rice gene given the sequence of an Arabidopsis gene which has been characterised already.
2- Comparing the similarity between two or more gene sequences to get a measure of their relatedness. Comparisons also allows taxonomy to be examined, as well as the drawing have been cloned and sequenced. ESTs can be aligned against that template with their overlapping ends to allow the construction of a contig, or contiguous sequence.
3- Design of primers for the PCR reaction. Online and offline tools and to have their computer design thousands of primers with little effort. Reconstruction genes from EST sequences (Expressed Sequence Tags are short pieces of genes which are expressed), and grouping of proteins into families. There are is a huge amount of work being undertaken to classify the proteins encoded by genes into superfamilies and families. The number of sequences in GenBank is very large, this would take a long time, so clever methods are used to model groups based upon smaller number of sequences. Pfam (Protein Families) is an extremely good example of the use of mathematics to group proteins into families and to make the information resulting from that both useful and accessible.

Major research efforts in the field include sequence alignment, gene finding, genome assembly, protein structure alignment, protein structure prediction of gene expression and protein-protein interactions, and the modeling of evolution.

5.5 Exponential growth of data banks:

Most of the people using this website (MolBiol.Net) will be students, graduate students, technicians, researchers, group leader, and other scientists. Some of the people who contribute to this site are bioinformaticians, those are the people that develop the tools and methods of analysis that the scientists use for data mining and analysis.

- 1981 creation at Los Alamos of GenBank: 270 sequences, 370.000 nucleotides.

Growth of GenBank (Millions sequence).

Evolution of computing power (evolution of the size of banks)

Source : http://www.ncbi.nlm.nih.gov/Genbank/genbankstars

5.5.1 NCBI (National Center for Biotechnology Information)

- National Institutes of Health
- National Library of Medicine

- Established in 1988 as a national resource for molecular biology information.
- NCBI creates public databases
- NCBI conducts research in computational biology
- NCBI develops software tools for analyzing genome data
- NCBI disseminates biomedical information

5.5.1.1 The NCBI web entry site provides a surface to enter various databases:

PubMed: allows access to the MEDLINE literature database

ENTREZ: search and retrieval system for:

- MEDLINE
- DNA sequence databases
- protein sequence databases
- PopSet: DNA sequences that have been collected to analyse the evolutionary relatedness of a population
- Genome: whole genomes of over 800 organisms
- OMOM
- Taxonamy
- Structure

BLAST: DNA and protein similarity search programs

OMIM: Online Mendelian Inheritance in Man: a catalog of human genes and genetic Disorders

Taxonomy: provides taxonomy / organism information

Structure: a database of macromolecular 3D structures, as well as tools for their visualization and comparative analysis.

Completed Genomes oct 02 (2002/2003, source EMBL, the European **M**olecular **B**iology **L**aboratory)

http://www.ebi.ac.uk/genomes/

- Archeas 16/16 (ex. *Pyrococcus abyssi*)
- Organelles : 291/308 (ex. *Homo sapiens* mitochondrion)
- Phages : 112/112 (ex. *Haemphilus influenzae* phage HP2)
- Plasmids : 279/280 (ex. *Escherichia coli* plasmid F)
- Viroids : 40/40 (ex. Peach Mosaic)
- Virus : 839/880 (ex. Hepatitis C virus, SARS virus)
- Bacteria : 76 (ex. *Helicobacter pylori*)
- Eukaryotes : 6/9 (ex. *Homo Sapiens*).

5.5.2 Nucleotide databases

- **GenBank**
- **dbEST:** **D**atabase for **E**xpressed **S**equence **T**ags
- **dbGSS**: **G**enome **S**urvey **S**equences **D**atabase
- **dbSTS:** **D**atabase for **S**equence **T**agged **S**ites
- **dbSNP:** **D**atabase for **S**ingle **N**ucleotide **P**olymorphism

5.5.2.1 GenBank

- GenBank is the NIH genetic sequence database, an annotated collection of all publicly available DNA sequences.
- GenBank (at NCBI), together with the DNA DataBank of Japan (**DDBJ**) and the **E**uropean **M**olecular **B**iology **L**aboratory (**EMBL**) comprise the International Nucleotide Sequence Database Collaboration.
- These three organizations exchange data on a daily basis.
- GenBank grows at an exponential rate, with the number of nucleotides bases doubling approximately every 14 months.

5.5.2.2 dbEST

- dbEST is a division of GenBank that contains sequence data and other information on

"single-pass" cDNA sequence, or Expressed sequence Tags.

- dbEST contains data from a number of organisms.

Number of ESTs, summary by Organism – March 23, 2007

Homo sapiens (human)	7,957,500
Mus musculus + domesticus (mouse)	4,745,026
Rattus sp. (rat)	871,163
Glycine max (soybean)	371,817
Arabidopsis thaliana (thale cress)	2,136,000

The I.M.A.G.E. consortium (http://image.Llnl.gov)

- The I.M.A.G.E. Consortium was initiated in 1993 by four academic groups to perform an **I**ntegrated **M**olecular **A**nalysis of **G**enomes and them **E**xpression.
- Prepare and share high-quality, arrayed cDNA libraries.
- Place sequence, map, and expression data on the clones in these arrays into the public domain.
- The human and mouse genomes are the first to be studied.
- Recently, rat, zebrafish, and Xenopus have been added, and additional cDNA libraries from other species will follow over time.
- Sequences from I.M.A.G.E. cDNA clones are submitted immediately to dbEST.

5.5.2.3 dbGSS

- The GSS division of GenBank is similar to the EST division.
- However, in contrast to the EST division its sequences are genomic in origin, rather than cDNA (mRNA).

- The GSS division contains (but is not limited to) the following types of data:
  - random "single pass read" genome survey sequences.
  - cosmid/BAC/YAC end sequences
  - exon trapped genomic sequences
  - Alu PCR sequences

5.5.2.4 dbSTS

- dbSTS is an NCBI resource that contains sequence and mapping **data on short genomic landmark sequences or Sequence Tagged Sites.**

5.5.2.5 dbSNP

- **dbSNP** is an independent database and not a division of GenBank. dbSNP was launched in September of 1998.
- SNP stands for "**single nucleotide polymorphism**"
- **SNPs are the most common genetic variations** and occur once every 100 to 300 bases.
- **A key aspect of research in genetics is the association of sequence variation with heritable phenotypes.**
- **The SNP will accelerate the identification of disease genes by allowing researchers to look for associations between a disease and specific differences (SNPs) in population.**
- The database has been designed to accept several classes of genetic variaton:
  - SNPs
  - microsatellite repeats
  - small insertion / deletion polymorphisms
- Currently only *Homo sapiens* is represented in the database; however, the

database has been designed to accept mutation information from any species, not just *Homo sapiens*.

5.5.3 GEO: Gene Expression Omnibus

http://www.ncbi.nlm.nih.gov/geo/

- In order to **support the public use and dissemination of gene expression data**, NCBI has launched the Gene Expression Omnibus.
- **Goal:** to build a gene expression data repository and online resource for the retrieval of gene expression data from any organism or artificial source.
- Many types of gene expression data from plattform types such as spotted microarray, high-density oligonucleotide array (HDA), hybridization filter and **S**erial **A**nalysis of **G**ene **E**xpression (SAGE) data, will be accepted, accessioned, and archieved as a public data set.

5.5.3.1 How to retrieve a database entry via the NCBI entry site ?

A: Searching by accession number (or other unique identifier)

- Go to http://www.ncbi.nlm.nih.gov/
- Select: ENTREZ
- Select Nucleotide to retrieve a DNA or RNA sequence
- Enter Accession number (eg. AC005662) and press Go button
- The next window opens, showing the result of your query; if the accession number you entered does not qualify a real accession (e.g. AC005662) this will be indicated (*'No items found'*). Sometimes, however, more than one item is found.
- Select the item you are interested in, here select the AC005662 link. The database entry appears on your screen.

5.5.3.2 The database entry contains additional links.

5.5.3.2.1 Link organism:

Link to the NCBI Taxonomy Browser;

Gives taxonomy ID number, and additional species information;

From here it is also possible to access all sequences (DNA, protein) and protein structure data from this organism.

To achieve this:

- Please press the Lineage (abbreviated) button, which then changes to Lineage (full) and also shows a Get Sequences line.
- select whether you want to access nucleotide or protein sequences or structures (of proteins); the numbers in brackets indicate the total number of sequences or structures available for this organism.
- Select Get Sequences.
- The next window opens and shows links to all database entries, which can now be individually selected.

5.5.3.2.2 Link MEDLINE:

Links to the literature database MEDLINE;

Shows title, author list, and abstract, including MeSH terms of the paper that is mentioned in the entry (AC005662).

5.5.3.2.2.1 Link mRNA:

Shows the complete mRNA sequence (given as DNA sequence), here bases 1-3000 bp.

5.5.3.2.2.2 Link CDS:

Shows the coding sequence only (translated region is 2319 bp, from ATG to stop codon).

5.5.3.2.2.3 Link protein id:

Shows the protein translated from the nucleotide of AC005662.

5.5.4 Searching by subject search

- Go to: http://www.ncbi.nlm.nih.gov/
- Select: ENTREZ
- Select Nucleotide to retrieve a DNA or RNA sequence
- Select protein retrieve a protein sequence

- Enter **search term** (e.g. *Trifolium repens*) and press Go button
- The next window opens, showing the result of your query.
- Database entries identified by your search term are presented; the accession numbers and

  short descriptions of the individual entries are shown.
- Note your search term (here '*Trifolium repens*') will identify all database entries

  containing the search term, i.e. database entries containing phrases like 'similar to search

  term', or 'specific-host=" search term" will also be displayed on your screen.

5.5.5 Searching by authors

- Enter author names in the format: last name plus initials (e.g., Abdel-Haliem, M.E.F).
- Do not use punctuation.
- This format instructs Entrez to search only the author field.
- Entrez automatically truncates on the author's name to account for varying initials and

  designations such as 2nd.
- If only a last name is used in the query box (e.g., Abdel-Haliem), Entrez will search all fields for that term.

5.5.6 Truncations

- Place an asterisk (*) at the end of a search term to find all records with a term that begins with that text string.
- For example, the truncated search term "immunoglob*" will retrieve all records in the database that contain the word immunoglobulin, immunoglobulins, immunoglobin, and immunoglobins.
- Entrez searches the first 150 variations of a truncated term.
- If a truncated term produces more than 150 variations, which is possible with terms like "bact*" Entrez gives the following warning: "Wildcard search for

'bact*' used only the first 150 variations. Lengthen the root word to search for all endings."

- Left-handed truncation is not possible (e.g., "*bacterium").

5.5.7 Boolean Operators

- **AND:** instructs Entrez to find all documents that contain BOTH terms.
- **OR:** instructs Entrez to find all documents that contain EITHER term.
- **NOT:** instructs Entrez to find all documents that contain search term 1 BUT NOT search term 2.
- Boolean operators, AND, OR, NOT must be entered in UPPERCASE (e.g., promoters OR response elements).

5.5.8 How to proceed when too many entries are identified by your search term?

- Go to: http://www.ncbi.nlm.nih.gov/
- Select: ENTREZ
- Select Nucleotide to retrieve a DNA or RNA sequence
- Select protein to retrieve a protein sequence
- Enter **search term** (e.g. *Arabidopsis thaliana*) and press Go button
- The next window opens, showing the result of your query
- The search term '*Arabidopsis thaliana*' will identify more than 190.000 database entries.
- Reduce the number of resulting search items by using the **limits** feature.
    - Example: select 'mRNA', only from 'GenBank', 'Publication Date', enter e.g. from '2001/01/01' to 2001/01/31', press Go button.
    - Result: only 195 entries will be identified.
    - Note: limiting parameters are given on top of the page.

5.5.9 Combining previous searching

- Select: History: a summary of your previous database searches will be given.
- Enter new search parameters in the query field, such as '# 1 AND #5', or '#3 AND starch' or similar, and select Go button or press Enter

- Results will be displayed.

5.5.10 BLAST

- BLAST = **B**asic **L**ocal **A**lignment **S**earch **T**ool
- BLAST is a set of **similarity search programs**
- BLAST is designed to explore all of the available sequence databases regardless of whether the query is protein or DNA.
- The BLAST programs have been designed for speed, with a minimal sacrifice of sensitivity to distant sequence relationships.
- The scores assigned in a BLAST search have a well-defined statistical interpretation, making real matches easier to distinguish from random background hits.
- BLAST uses a heuristic **algorithm which seeks local** as opposed to global alignments and is therefore able to detect relationships among sequences which share only isolated regions of similariry (Altschul *et al.*, 1990).
- The BLAST Tutorial:

http://www.ncbi.nlm.nih.gov/Education/blasttutorial.html

5.5.10.1 Nucleotide BLAST

Nucleotide BLAST searches allow one to input nucleotide sequences and compare these against othe nucleotides.

1. **Standard nucleotide-nucleotide BLAST:** Takes nucleotides sequences in FASTA format, GenBank Accession numbers or G1 numbers and compares them against the NCBI nucleotide databases.
2. **MEGABLAST:** This program uses a 'greedy algorithm' (Webb Miller *et al.*,) for nucleotide sequence alignment searches and concatenates many queries to save time spent scanning the database. It is optimised for aligning sequences that differ slightly and is up to 10 times faster than more common sequence similarity programs. It can be used to swiftly compare two large sets of sequences against each other.
3.

3. **Search for short, nearly exact sequences**

- This search is similar to the standard nucleotide-nucleotide BLAST with the parameters set automatically to optimise for searching with short sequences.
- A short query is more likely to occur by chance in the database. Therefore, increasing the Expect value threshold, and also lowering the word size is often necessary before results can be returned.
- Low Complexity filtering has also been removed since this filters out larger percent- age of a short sequence, resulting in little or no query sequence remaining.

5.5.10.2 The BLAST Search box

- The BLAST 'Search' box accepts a number of different types of input
- it automatically determines the format
- Accepted input types are:

FASTA:

- A sequence in FASTA format begins with a single-line description, followed by lines of sequence data.
- The description line is distinguished from the sequence data by a greater-than (">") symbol in the first column.

Bare Sequence:

- This may be just lines of sequence data, without the FASTA definition line.
- It can also be sequence interspersed with numbers and/or spaces, such as the sequence portion of a GenBank/GenPept flatfile report.

Identifiers

- Normally these are simply accession, accession version or gi's, but a bar-separated NCBI sequence identifier will be accepted.
- These NCBI sequence identifiers have a very specific syntax.

5.5.10.2.1 Nucleotide Sequence Databases that can be used for searches

- **nr**

All GenBank + EMBL + DDBJ + PDB sequences (but not EST, STS or

GSS). Where: DDBJ is DNA Data Bank of Japan, STS is single nucleotide.

- **month**

  All new or revised GenBank + EMBL + DDBJ + PDB sequences released in the last 30 days.

- **dbest (Database for Expressed Sequence Tags)**

  Database from GenBank + EMBL + DDBJ sequences from EST Divisions.

- **Yeast**

  Yeast (*Saccharomyces cerevisiae*) genomic nucleotide sequences.

- ***E. coli***

  Escherichia coli genomic nucleotide sequences.

- **vector**

  Vector subset of GenBank

- **mito**

  Database of mitochondrial sequences

- **epd**

  Eukaryotic Promotor Database found on the web at http://www.genome.ad.jp/dbget- bin/ bfind?epd (KEGG: Kyoto Encyclopedia of Genes and Genomes).

5.5.10.2.2 How to perform a BLAST search?

- Paste query sequence into Search box.
- Select sub-sequence (if wanted).
- Chose database that you want the check for sequences homologous to your query.
- Press 'BLAST' button
- The next www page entitled '**formating BLAST**' opens, indicating the length (bp) of

  your query sequence, a Request ID (request identification number).
- Press 'Format' button to check results of BLAST search; note: a new www window

opens here, entitled '**results of BLAST**'.

- 'Taxonomy Reports': provides taxonomy information about the organisms that have sequences homologous to your query sequence.
- **Expect**
    - A statistical significance threshold for reporting matches against database sequences.
    - The default value is 10, this means that 10 matches are expected to be found merely by chance.
    - Lower expect thresholds are more stringent, leading to fewer chance matches being reported. Increasing the threshold shows less stringent matches. Fractional values are acceptable.

Results from BLAST searches are stored for some time. They can be retrieved from the NCBI database with the help of the request ID number.

Search for short, nearly exact sequences, much higher expect values are given here.

5.5.10.2. 3 Pairwise BLAST

- Allows to **BLAST 2 sequences directly against each other**.
- Example: compare sequences from AC005662 and AL138655.
- Provides a **graphical output** of the aligned sequences
- Provides a **sequence alignment.**

pairwise comparison of the sequences of the two promoters a 150 bp region was found that exhibited 83% similarity from the whole promoters. (Positions 886 to 1042 bp in AL138655 and 996 to 1154 bp in AC005662).

```
AC..:996 agagggaagcaaggagatgctaaatatagagaattcgagggtcttgcttgtttctctaaa 1056
         ||||| ||||  |||||||||||  ||||||||     ||||| ||||||||||||||||
AL..:886 agaggaaagctcggagatgctaagaatagagaaaaat-gggtcatgcttgtttctctaaa 944

AC..:1057 tttcaaggctttcattggtttgaggatatgaacgagtcctttctttgaatgacaaaatta 1116
          ||| |||||||||||||||||||||| ||||||| ||   ||||||||| |||| |||||
AL..:945  tttgaaggctttcattggtttgaggagatgaacg-gtttcttctttgaacgacataatta 1003

AC..:1117 gtttcta-aaactgataagaaaagatgaaatctcggaga 1154
          |||   |  || ||| ||||||||||||||| |||||||
AL..:1004 gttaagatgaattgacaagaaaagatgaaatgtcggaga 1042
```

• **Sequence comparison is a basic operation in molecular biology**

• It occurs, for instance, in:

- Search in Data Banks
- Genome Assembly
- Phylogenic trees construction

• The knowledge of principles of sequence comparison is necessary for mastering the interpretation of results given by softwares.

5.5.11 Protein Sequence Databases

5.5.11.1 The SWISS-PROT Database

• The SWISS-PROT database consists of sequence entries. It contains high-quality annotation, is non-redundant and cross-referenced to many othe databases.

• SWISS-PROT is accompanied by TrEMBL

• TrEMBL a computer-annotated supplement to SWISS-PROT. TrEMBL contains the translations of all coding sequences (CDS) present in the EMBL Nucleotide Sequence Database.

• SWISS-PROT is an annotated protein sequence database.

• It was established in 1986 and maintained collaboratively, since 1987, by the **S**wiss **I**nstitute of **B**ioinformatics (**SIB**) and the **E**uropean **B**ioinformatics **I**nstitute (**EBI**).

• The SWISS-PROT protein sequence database consists of sequence entries. Sequence entries are composed of different line types, each with their own format. For standardization purposes the format of SWISS-PROT follows as closely as possible that of the EMBL.

5.5.11.2 Nucleotide Sequence Database

- The SWISS-PROT database distinguishes itself from other protein sequence databases by four distinct criteria: Annotation in SWISS-PROT, as in most other sequence databases, two classes of data can be distinguished: the core data and the annotation.
- For each sequence entry the core data consists of: the sequence data; the citation information (bibliographical references); the taxonomic data (description of the biological source of the protein).

The annotation consists of the description of the following items:

- **Function** (s) of the protein
- **Post-translational modification** (s) (for example carbohydrates, phosphorylation, etc.)
- **Domains and sites** (for example zinc fingers, calcium binding regions, SH2 domain, etc.)
- **Secondary structure** (for example α helix, β sheet, etc.)
- **Quaternary structure** (for example homodimer, heterotrimer, etc.)
- **Disease** (s) associated with deficiencies in the protein.
- Annotation is mainly found in the comment lines (CC), in the feature table (FT) and in the keyword lines (KW).
- SWISS-PROT is currently cross-referenced with about 30 different databases.
- SWISS-PROT documens-user manual, release notes, indices and lots of other important documents and lists.
- **Database access:**
- The ExPASy in Geneva offers the choice of full-text search or of individual lines (e.g. ID, AC, DE, OS, OG, GN, RL, RA)
- SRS for more complex and/or multiple database queries.

ExPASy Molecular Biology Server

http://www.expasy.ch/

• ExPASy = **Ex**pert **P**rotein **A**nalysis **S**ystem proteomic server of the Swiss Institute of Bioinformatics (SIB).

Where:

• SRS access to SWISS-PROT, TrEMBL and other databases using the Sequence Retrieval System.

• by accession number or ID (AC or ID line; SWISS-PROT and TrEMBL)

• by description or identification (any word in the DE, OS, OG, GN and ID lines; SWISS-PROT and TrEMBL).

• by author (RA line; SWISS-PROT and TrEMBL).

• by citation (RL line; SWISS-PROT only).

5.5.12 Human Proteomics Initiative

http://www.expasy.ch/sprot/hpi/

The **H**uman **P**roteomics **I**nitiative (HPI) is a major project to annotate all known human sequences according to the quality standards of SWISS-PROT. This means providing, for each known protein, a wealth of information that include the description of its function, its domain structure, subcellular location, post-translational modifications, variants, similarities to other proteins, etc.

http://www.expasy.ch/sprot/hamap/

The HAMAP project aims to automatically annotate a significant percentage of proteins originating from microbial genome sequencing projects.

PROSITE

Database of protein families and domains

http://www.expasy.ch/prosite/

• PROSIT E is a database of protein families and domains.

• It consists of biologically significant sites, patterns and profiles that help to reliably identify to which known protein family.

• PROSITE is a method of determining what is th function of uncharacterized proteins translated from genomic or cDNA sequences.

• It consists of a database of biologically significant sites and patterns, but in some cases the sequence of an unknown protein is too distantly related to any protein of known structure to detect its resemblance by overall sequence alignment, but it can be identified by the occurrence in its sequence of a motif, signature, or fingerprint.

5.5.13 Related services

• **The Pfam Database**

http://www.sanger.ac.uk/Pfam/

• **Pfam** from the Sanger centre in Hinxton (UK) or from Washington University (USA). Pfam has very accurate descriptions of protein domains. There are basically two ways

you will want to use this information

- Using an established SWISSPROT sequence where the domain structure of the protein in the data base of computer.
- Using a completely new protein sequence and asking what Pfam thinks the domain structure.

• **BLOCKS**, this is based around automatic ungapped alignments

• **PRODOM**, this is an automatically generated domain database based on PSI-BLAST searching.

• **PRINTS**, this is based around protein "finger-prints" of a series of small conserved motifs making up a domain.

• **SMART**, this is a database concentrating on extracellular modules and signaling domains.

5.5.14 The InterPro Database

http://www.ebi.ac.uk/interpro/

InterPro: (**I**ntegrated **R**esource of **P**rotein **F**amilies, **D**omains and **F**unctional **S**ites).

- Unfortunately, the various signature databases (such as PROSITE, ProDom etc.), do not share the same formats and nomenclature, and each database has is own strengths and weaknesses.
- To capitalise on these, the following partners: EBI, SIB, University of Manchester, Sanger Centre, GENE-IT, CNRS/INRA, LION bioscience AG and University of Bergen **unified PROSITE, PRINTS, ProDom and Pfam into InterPro**.
- InterPro provides an integrated view of the commonly used signature databases, and has an intuitive interface for text and sequence-based searches.

How to perform an InterPro Scan?

- Example: InterPro scan with Swiss-Prot entry P20115
- Select 'Sequence Search' option on InterPro entry page.
- Enter protein sequence.
- Enter your email address.
- Submit job.
- Results will be displayed.

5.5.14.1 InterPro Member Databases

• **The SWISS-PROT** database consists of sequence entries. It contains high-quality annotation, is non-redundant and cross-referenced to many other databases. SWISS-PROT is accompanied by TrEMBL, a computer-annotated supplement to SWISS-PROT. TrEMBL contains the translations of all coding sequences (CDS) present in the EMBL Nucleotide Sequence Database, which are not yet integrated into SWISS-PROT.

• **PROSITE** is a database of protein families and domains. It consists of biologically significant sites, patterns and profiles that help to reliably identify to which known protein family a new sequence belongs.

• **Pfam** is a large collection of multiple sequence alignments and hidden Markov models covering many common protein domains.

• **PRINTS** is a compendium of protein fingerprints. A fingerprint is a group of conserved motifs used to characterise a protein family; its diagnostic power is refined by iterative scanning of a composite of SWISS-PROT + SP-TrEMBL usually the motifs do not overlap, but are separated along a sequence, through they may be contiguous in 3D-space. Fingerprints can encode protein folds and functionalities more flexibly and powerfully than can single motifs, their full diagnostic potency deriving from the mutual context afforded by motif neighbours.

• **The PreDom** protein domain database consists of an automatic compilation of homologous domains. Current versions of ProDom are built using a novel procedure based on recurvsive.

• **SMART** (a Simple Modular Architecture Research Tool) allows the identification and annotation of genetically mobile domains and the analysis of domain architectures. More than 500 domain families found in signalling, extracellular and chromatin-associated proteins are detectable. User interfaces to this database allow searches for proteins containing specific combinations of domains in defined taxa.

5.5.15 BCM Search Launcher (Multiple Sequence Alignments)

> 1st gene

> 2nd gene

>3rd gene

> 4th gene

then alignment→ appear blue box and other, take only blue colour→ Box schades (strg + c) → RTF-new r → strg + v → Runbox schade.

5.5.15.1 Sequence analysis of the coding regions of both *AtPIP5K-f* and *AtPIP5K-x*

Amino acid sequence analysis indicated that the proteins AtPIP5K-f and x clearly belonged in the PIP5K family of enzymes. The AtPIP5K-f protein shares 86% identical amino acid residues with AtPIP5K-x protein. The comparison of the amino acid sequences of AtPIP5K-f, x and Mss4p are 36% identical (195 residues), with an additional 127 residues conserved between them. Searching the sequence data bases with the BLAST algorithm, significant similarity to two *S. cerevisiae* gene products was found. These were Fab1p and Mss4p at their C-terminal ends within this region, the identity between AtPIP5Kf, x and these proteins is 30% and 36% respectively. If the conservation amino acid replacements are taken into account, the similarity among the four proteins is close to 53% in this region (Fig. 5.1). No significant similarity between Fab1p and Mss4p can be detected outside this catalytic domain, with the unique nature of the remaining sequences for each postulated to confer specificity of function.

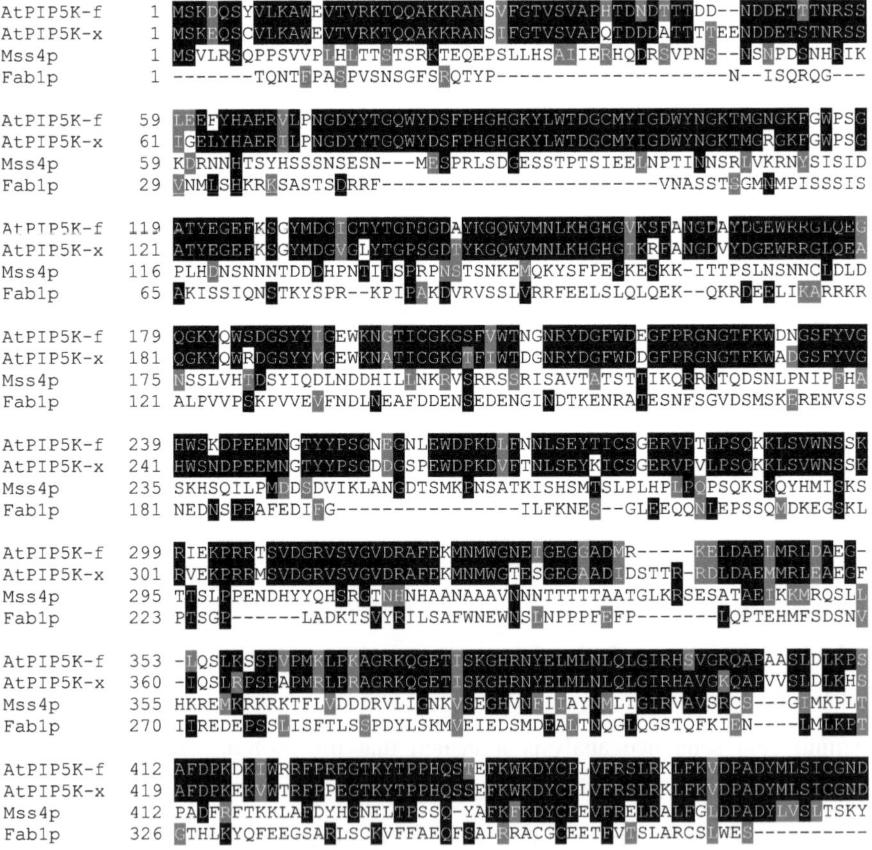

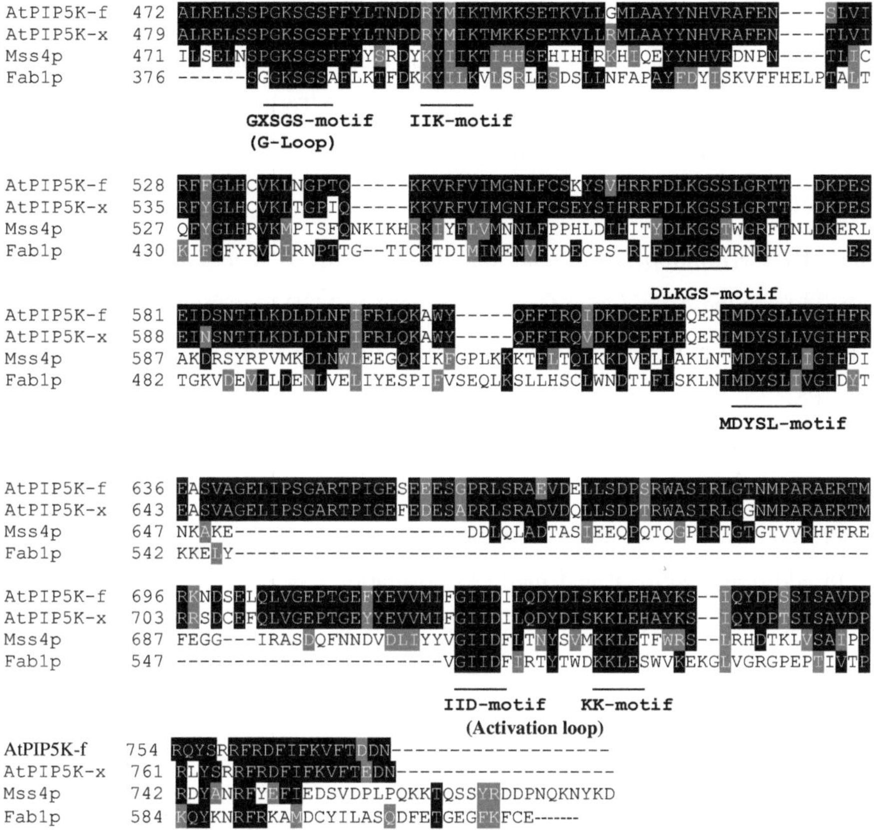

Fig. 5.1: Alignment of the amino acid sequences of AtPIP5K-f and AtPIP5K-x with those of *S. cerevisiae* Fab1p and Mss4p. Alignment of the amino acid sequence was established by using BCM Search Launcher, the homologies are shown by shading. Residues conserved in all four sequences, constituting elements of the kinase domain, are underlined. GXSGS-motif (G-Loop) resembles the phosphate-binding loop of protein kinases, heterotrimeric G-proteins; IIK-motif coordinates with the α-phosphate of ATP; the (Y/F) DLKGS-motif may be functionally analogous to the catalytic loop and magnesium coordination residues of protein kinase; a MDYSL-motif and an IID-motif are corresponding residues in protein kinase A.

5.5.16 The MIPS Database

http://mips.gsf.de/

MIPS is the **M**unich **I**nstitute for **P**rotein **S**equences. MIPS is a member of PIR-International (**P**rotein **I**dentification **R**esource) and of EMBNET (**E**uropean **M**olecular **B**iological **N**etwork).

Yeast (*Saccharomyces cerevisiae*)

The MIPS **Y**east **G**enome **D**atabase (MYGD) presents a comprehensive database which summarizes the current knowledge regarding the more than 6000 ORFs encodeed by the Yeast Genome. Search Tools (for Gene Names, Codes and Synonyms, for Text Strings).

Arabidopsis thaliana

The **M**IPS *Arabidopsis thaliana* **d**ata**b**ase (MAtDB), contains all sequences derived from the **A**rabidopsis **G**enome **I**nitiative (AGI) sequencing project.

- Databases of gDNA and protein sequences from ESSA project are maintained at MIPS.
- Beside the opportunity to do BLAST similarity searches against these databses, MAtDB contains a large set of data on individual genes and clones which can be accessed via a searchable index.
- In addition positional and marker information can be used to zoom from a chromosomal, via distinct regions to single BAC clones and finally to single genes.

To Know Atg

- MIPS *Arabidopsis thaliana* database is the www access to data of the *Arabidopsis* Genome Initiative compiled, analysed. Annotated and stored at MIPS by the MIPS *Arabidopsis* group.

// mips.gsf.de

↓search

(google for *Arabidopsis* of chromosome) → no. Of Bac.

DNaster

↓

then Edit sequence → Map Draw → open the sequence.

5.5.16.1 The MAtDB web page

Interactive Redundancy Viewer

- The interactive Redundancy Viewer is a Java applet which allows you to explore redundancies in the Arabidopsis thaliana Genome on your own.

• You can zoom into areas of interest, modify display settings and link to protein reports.

• Consider that quite a lot of data have to be transferred, so a good internet connection might be advantageous.

Protein analysis

• Protein Function Analysis / InterPro Analysis

• List of proteins with homology to human disease genes

• Fasta Database (all Arabidopsis proteins tested against all other proteins in the database).

Chromosome analysis

• Example: Graphical View/ Chromosome 1: provides you with a graphical presentation of chromosome 1 (covering 30 million bp in total).

• Select a region of chromosome 1: the next view presents the BAC clones that cover the chromosome over 2 Mio. Bp

• You can 'walk' along the chromosome by clicking the >> or << symbols.

• Click on an individual BAC to get information about all the predicted genes in this region.

5.5. 16.2 The Arabidopsis Information Resource (tair)

http://www.arabidopsis.org/

• PatMatch: allows to identify proteins by searching with short peptide sequences; also allows to identify DNA sequences by searching with short DNA stretches (e.g. if one wants to identify *cis*-elements in promoter regions).

www.arabidopsis.org → restriction analysis (site map)→ restriction enzyme.

http://arabidopsis.org/servlets/search

http://arabidopsis.org/info/expression

Pattern Search is possible at:

• Using PatMatch at www.arabidopsis.org

• Using PEDANT at pedant.gsf.de

• *At* MatDB: http://mips.gsf.de/proj/thal/db

5.5.16.3 The TIGR Database

http://www.tigr.org/

- TIGR is **T**he **I**nstitute for **G**enome **R**esearch
- The TIGR Databases are a collection of curated databases containing DNA and protein sequenc (microbial Database, Rice database, *Arabidopsis thaliana*), gene expression, cellular role, protein family, maintains an alternative *Arabidopsis* genome database and taxonomic data for microbes, plants and humans.

5.5.16.4 Enzyme cutter

http://tool.neb.com/NEBcutter/index.php3

rebase (site map) → ecori (small letter)

http://rebase.neb.com/rebase

www.molecular cloning.com (Mannitus Mannual)

www.r9corporation.fsnet.co.uk (Bioinformatic and Molecular Biology).

Expasy → DNA-protein → proteomic Tools → translate sequence.

prettyseq → output sequence with translated ranges (EMBOSS)

http://bioweb.pasteur.fr/sequnal/interfaces/prettyseq.html

5.5.16.5 Promoter analysis

Transfac

http://oberon.rug.ac.be:8080/plantCARE/index.html

ZKBS vector list (Die Zentrale Kommision fuer Biologische Sicherheit)

http://www.rki.de/GENTEC/ZKBS/

T-DNA Insertion (knock out)

http://nasc.nott.ac.uk/insertwatch

Salk Institute with GABI-Kat

http://signal.salk.edu/cgi-bin/tdna express

https://www.mpiz.koeln.mpg.de/~GABI-Kat/home/index.html

https://www.mpiz.koeln.mpg.de/~GABI-Kat/MASC-SNPS

***Arabidopsis* membrane protein library (AMPL)**

http://www.cbs.umn.edu/arabidopsis

5.5.17 Program Rice Genome Research (PRGR)

http://www.nias.offrc.go.jp/project/integenome/index.html

5.5.18 Rice membrane proteins library (RMPL)

http://www.cbs.umn.edu/rice

http://www.knowledgebank.irri.org/ (Rice Knowledge Bank)

http://www.harvestplus.org/ (Breeding Crops for Better Nutrition).

5.5.19 Metabolic / Enzyme databases

5.5.19.1 BRENDA : Enzyme Information Service

http://www.brenda.uni-koeln.de/

- BRENDA is the main collection of enzyme functional data available to the scientific community.
- The enzymes are classified according to the Enzyme Commission list of enzymes. Some 3500 "different" enzymes are covered. Frequently enzymes with very different properties are included under the same EC number.
- BRENDA is maintained and developed at the institute of Biochemistry at the University of Cologne. Data on enzyme function are extracted directly from primary literature by scientists holding a degree in Biology or chemistry.
- The data collection is being developed into a metabolic network information system with links to Enzyme expression and regulation information.
- The development of an enzyme data information system was started in 1987 at the German National Research Centre for Biotechnology in Braunschweig (GBF) and is now continued at the University of cologne, Institute of Biochemistry.
- Links is provided to: SwissProt.
- Link is provided to KEGG.

5.5.20 KEGG: Kyoto Encyclopedia of Genes and Genomes

http://www.genome.ad.jp/kegg/

• KEGG was initiated in May 1995 under the Japanese Human Genome Program.

• The primary objective of KEGG is to computerize the current knowledge of molecular interactions; namely, metabolic pathways, regulatory pathways, and molecular assemblies.

• KEGG also started to organizes a database of all chemical compounds in living cells, maintains gene catalogs for all the organisms that have been sequenced and links each gene product to a component on the pathway.

• KEGG also aims at developing new bioinformatics technologies toward functional reconstruction.

5.5.20.1 EMP: Enzymes and metabolic pathways

http://emp.mcs.anl.gov/

• EMP is a comprehensive electronic source of biochemical data.

• It covers many aspects of enzymology and metabolism

• The database format has about 300 subject fields (e.g. metabolism, enzyme and reaction, enzyme assay and purification, and immunochemistry, etc.,)

• EMP contains more than 3,000 metabolic diagrams.

5.5.20.2 WIT: What is there?

5.5.20.2.1 Interactive Metabolic Reconstruction on the Web

http://wit.mcs.anl.gov/WIT/

• WIT is a www-based system to support the curation of function assignments made to genes and the development of metabolic models.

• What is a Metabolic Reconstructiob? A metabolic reconstruction is a model of the metabolism of the organism derived from sequence, biochemical, and phenotypic data.

5.5.20.2.2 Boehringer Mannheim-Biochemical Pathways

http://www.expasy.ch/cgi-bin/search-biochem-index

• Metabolic pathways.

• Cellular and Molecular Processes.

5.5.21 Yeast Pathways in MIPS

http://www.mips.biochem.mpg.de/proj/yeast/pathways/index.html

• select 'Pathways'

5.5.22 Human database

5.5.22.1 The Genome Database (GDB)

http://www.gdb.org/

• Established at Johns Hopkins University in Baltimore, Maryland, USA in 1990.

• The Genome Database (GDB) is the official central repository for genomic mapping data resulting from the Human Genome Initiative.

• GDB's mission is to make available to scientists an encyclopedia of the human genome.

• At present, GDB comprises descriptions on the following types of objects:

- Regions of the human genome, including genes, clones, amplimers (PCR markers), breakpoints, cytogenetic markers, fragile sites, ESTs, syndromic regions, contigs and repeats.
- Maps of the human genome, including cytogenetic maps, linkage maps, radiation hybrid maps, content contig maps, and integrated maps. These maps can be displayed graphically via the Web.
- Variations within the human genome including mutations and polymorphisms, plus allele frequency data.

5.5.22.2 GeneCards

http://nciarray.nci.nih.gov/cards/

• GeneCards™ is a database of human genes, their products and their involvement in diseases.

• To test new approaches for the efficient navigation of biomedical information, the GeneCards Encyclopedia has been designed.

• A crucial aspect of the Gene Cards strategy: to make use of standard nomenclature.

- See HUGO Gene Nomenclature Committee at
  http://www.gene.ucl.ac.uk/nomenclature/
- See Guidelines for Human Gene Nomenclature (1997) at
  http://www.gene.ucl.ac.uk/nomenclature/guidelines.html

5.5.22.3 HUGE: Database of **H**uman **U**nidentified **G**ene-**E**ncoded Large Proteins

http://www.kazusa.or.jp/huge/

- **Human cDNA project at the Kazusa DNA Research Institute.**
- In this project, it is planned to sequence and analyze long (>4 kb) human cDNAs and to establish methods by using the sequence data how to predict the primary structure of proteins of various biological activities. Currently, the focus is on the analysis of cDNA clones encoding particularly large proteins (>50 kDa).
- The HUGE protein database contains various types of information derived from the predicted primary structure data of newly identified human proteins, and also cDNA sequences, restriction maps, homology data, Motif / Profile / Pfam data, expression pattern (Northern Blot), a link to GeneCards are provided.
- Homology search against HUGE database is also possible.

5.5.22.4 Mitomap: A Human Mitochondrial Genome Database

http://www.gen.emory.edu/mitomap.html

- a summary of polymorphisms and mutations of the human mitochomdrial DNA

5.5.22.5 The Whole Mouse Catalog

http://www.rodentia.com/wmc/index.html

5.5.22.6 Bovine Genome Database

http://bos.cvm.tamu.edu/bovgbase.html

euGenes

http://iubio.bio.indiana.edu:8089

- Genomic information for eukaryotic organisms

5.5.22.7 The WWW Virtual Library: Model Organisms

http://ceolas.org/VL/mo/#arabidopsis

ARS Genome Database Resource

http://ars-genome.cornell.edu/index.html

- ARS Genome Database Resource is primarily a site for communication of plant Genome data among scientists worlwide. The medium is a set of genome databases, similar to the ones for Human genome mapping and sequencing.

5.5.22.8 How to submission of sequence in GenBank?

- Open NCBI, and look under sequence submission → left side → Bankt→ sequence here without stop codon.→ press.

5.5.22.9 How to draw the family tree?

You have two possibility (two programs):
1. Clustalx.exe (this program load from internet)

- File (load sequence in Fasta text)

- Alignment (do complete alignment)

- Draw the tree (2 kinds 1. ph. and 2. ph. b)

Tree view from internet

	→ ph.
File (open)	
	→ ph. b

Then with scale (0.1) is the best.

The second program by using e. mail

2. BiBisew (DIALAGen)

- submission protein sequence (> Fasta)
- e. mail address
- then open your e. mail address, take the last things matrix
- Then open phyL I (2nd page phyL I)
- Consensus trees (draw trees, the best your matrix unrooted, phylogenesis submitted accept).

6. Transgenic Organisms

In order to express foreign genes in plants or animals the coding sequences of the gene in question has to be inserted into an expression cassette. This cassette should provide promoter and terminator sequences. A number of techniques for manipulating DNA. One of these was the method for taking a piece of DNA from any source, and inserting it into a bacterial cell, in other words cloning it. The recipient bacteria may or may not express the protein the gene codes for, but the method of getting "foreign" DNA into a bacterial cell was essentially rather simple. Bacteria with the foreign DNA are said to have been transformed: their genome has been changed; new genetic characters have been inserted. Such bacteria may then be said to be transgenic, and it soon became the goal to make transgenic animals and plants. This was important as it would allow genes to be transferred across the species barrier in eukaryotes, a given gene from any organism. This works because the genetic code is universal.

6.1 Hybridization probe (random prime labelling system)

Nucleic acid hybridization can be used to identify a particular recombinant clone if a DNA or RNA probe, complementary to the desired gene. Rediprime allows DNA from a variety of sources to be labelled to high specific activity using (^{32}P) dCTP. The system has been designed for use with Redivue (^{32}P) dCTP with a specific activity of 3000Ci/mmol. Each reaction tube can label up to 25 ng of DNA and after incubation for 10 minutes at 37°C, probes with a specific activity of 1.9 x 10^9 dpm/μg or greater can be produced.

6.2 Blotting (RNA or DNA)

The RNA or DNA was blotted to nylon membranes (Hybond N, Amersham Pharmacia Biotech) overnight (Fig. 6.1), and Fixed the nucleic acid to the membrane by baking at 80°C for 2 hours or by using an optimized UV crossliniking procedure (Stratalinker, function: "autocrosslink").

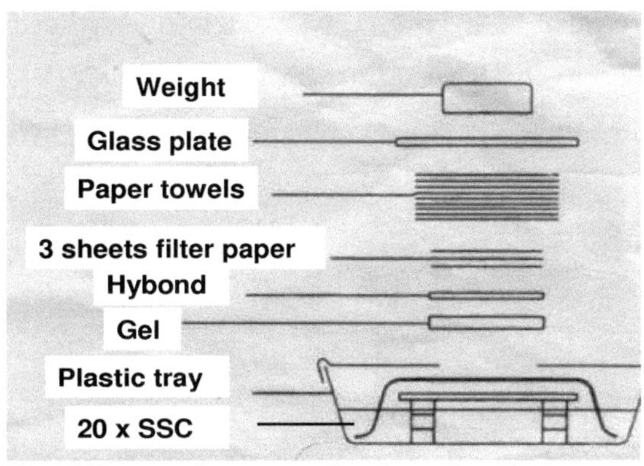

Fig. 6.1 Digrammatic representation of a capillary blotting

6.3 Northern blotting

30 μg total RNA was separated on a 1.5% denaturing agarose gel containing formaldehyde (Lehrach *et al.*, 1977). The RNA was blotted to nylon membranes (Stratalinker, function: "autocrosslink"). The membranes were prehybridized at 60°C for 2 hours in 250 mM sodium phosphate buffer (PH 7.2) containing 7% SDS, 1 mM EDTA and 1% BSA. Hybridization of the membranes was performed in the same buffer at 60°C overnight with a radioactive labelled probe (Rediprime II DNA labelling system). The membranes were washed 15 min. with 5x SSC and 0.5% SDS, 10 min. with 1x SSC and 0.5% SDS and 5 min. with 0.2x SSC and 0.5% SDS. The labelled RNA on the membrane was visualized by autoradiography using x-ray films (Biomax MS, Eastman-Kodak, Stuttgart, Germany) developed at –80°C over three to five days. The following radioactively labelled probes were used.

Gene expression is often cell-specific, mRNA transcripts are often analyzed by a blotting technique called Northern blotting, the RNA samples are electrophoresed on agarose gels containing a denaturing reagent such as formaldehyde to disrupt intrastrand base pairing and RNA secondary structures. The RNA is transferred from the gel onto a solid support, such as nylon or nitrocellulose membrane, so that the pattern of the RNA in the gel is preserved on the membrane support as shown in Fig. 3.5. The filter membrane containing single strand RNA can then be hybridized with single-stranded cDNA probes that have been radioactively labeled in vitro using ^{32}P-dNTP and DNA polymerase. The washed filter is exposed to film for autoradiography.

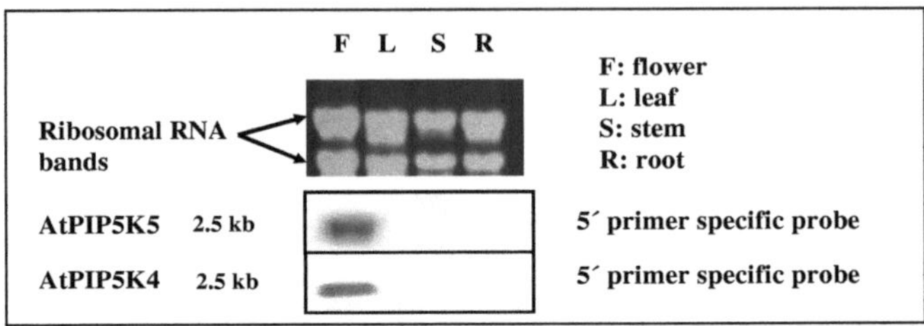

Fig. 6.2 Northern blotting to characterize RNA species in various cell types. The RNA is separated by gel electrophoresis in a denaturing formaldehyde agarose gel, transferred to a nitrocellulose filter, and hybrid with gene-specific cDNA probe. Note both genes are expressed in flower.

6.4 DNA preparation (genomic DNA from plant tissue)

Genomic DNA was prepared with cetyl-trimethylammoniumbromide (CTAB-extraction buffer)

1 take 20-40 µg of genomic DNA.

2 Add 1µl of enzyme + buffer overnight in the morning, and then after 6 hours add 1 µl enzyme.

3 Run 0.8 % agarose gel in 1x TAE + EtBr of Big gel, 100-120 V for 4h.

4 take 20-40 µg of genomic DNA.

5 Add 1µl of enzyme + buffer overnight in the morning, and then after 6 hour add 1 µl enzyme.

6 Run 0.8 % agarose gel in 1x TAE + EtBr of Big gel, 100-120 V for 4h.

7 Gel make photo, then 15 min in 0.25 M HCl (denaturation solution), rinse in water for 30 min., and then 30 min in neutralization solution.

6.5 Southern blot

The DNA was digested with a suitable restriction enzyme, electrophoresed and transfered to nitrocellulose membrane "Hybond N". The transfer was done using the capillar transfer alkali method in 0.4 N NaOH. The DNA was fixed on the membrane by 15s exposure to UV light. The filter was neutralized in 2xSSC.

Prehybridization, hybridization and detection were carried out using standard techniques for DIG nucleic acid labeling and detection systems. The dot blots from phages (plaques hybridization) were done using the same techiques. The phages (about 10^6 and 10^8 particles / ml) were transferred onto Hybond N filters for about 5 min. The filters were denatured in 0.4 N NaOH and then neutarlized in 6xSSC, 2xSSC. In the case of genomic DNA, the probe was radiolabelled with α- (^{32}P) dATP. After washing the filters were exposed to X-ray films (Fig. 6.3).

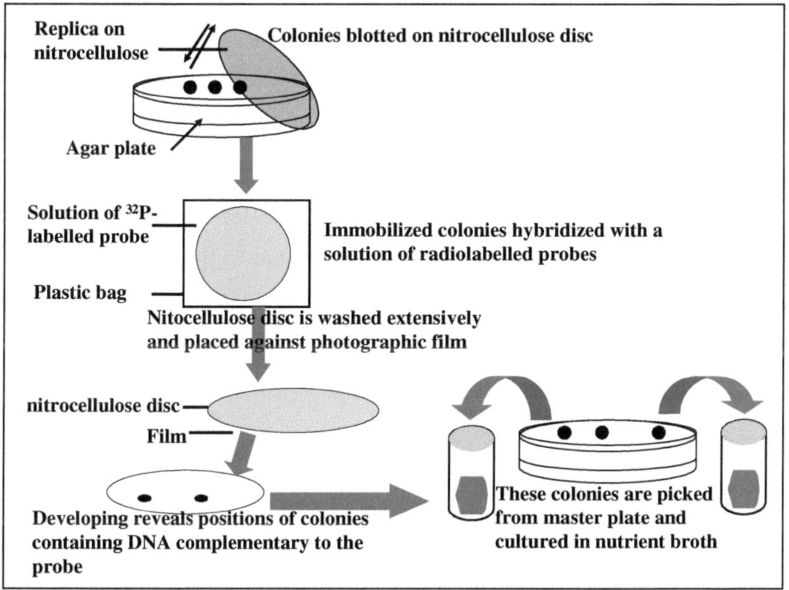

Fig. 6.3 Southern blot. A replica of the colonies on the plate is taken by blotting lightly on to a nitrocellulose disc. The "print" of colonies on the nitrocellulose is lysed to release the DNA and this is baked on to the sheet and made single stranded. Immersion in a solution of the radioactively labelled probes for the DNA (an oligo) allows this to hybridize with the DNA released from the colonies. If the nitrocellulose is now dried and placed against photographic film in the dark, on development black spots may be seen corresponding to colonies which contain the DNA. We can now go back to the original agar plate from which the blot was made and pick of some of the identified colonies. These may be grown up to produce lots of the DNA that has been cloned. This DNA is in the plasmids in the bacteria.

6.6 Protein extractions from plant leaves and seeds

The frozen plant materials were homogenized in liquid nitrogen and then extracted in extraction buffer A (50 mM Tris-HCl pH 6.8, 2 % Triton x 100 and

1 % β-mercaptoethanol) or extraction buffer B (phosphate buffered saline [PBS]). The protein extract from 100 mg plant material was stirred for 1h in 300 μl extraction buffer. The slurry was centrifuged (15,000 rpm, 15 min. at 4 °C) and supernatant used for enzyme activity assay and Western blot (Sambrook *et al.*, 1989).

6.7 SDS-polyacrylamide gel electrophoresis of proteins (PAGE)

Amost all analytical electrophoresis of proteins is carried out in polyacrylamide gels under conditions that ensure dissociation of the proteins into their individual polypeptide subunits and that minimize aggregation. Most commonly, the strongly anionic detergent SDS is used in combination with a reducing agent and heat to dissociate the proteins before they are loaded on the gel. The denatured polypeptides bind SDS and become negatively charged. Because the amount of SDS bound is almost always proportional to the molecular weight of the polypeptide and is independent of its sequence, SDS-polypeptide complexes migrate through polyacrylamide gels in accordance with the size of the polypeptide. At saturation, approximately 1.4 g of detergent is bound per gram of polypeptide.

6.8 Protocol of Western blot

About 5 μg of probe protein with Laemmli buffer (~ 5 μl)

→ Laemmli (2x) before used with β-Mercapto-ethanol (dilution 9:1)

Then denaturation 10 min at 95 °C

Gel in comb and filled with running buffer (1x)

Load with probe, gel running with 120-130 V ~ 2h.

6.9 Western blot

1. Moisten the Grafitplate with 1x transfer buffer, the moisten the site with 3 MM-paper, and then the protein gel, and the plate with negative pole.
2. Above the 3 Whatman MM-paper come the gel after put in the transfer buffer (This important!!!)

3. Nitrocellouose or PVDF membrane (moisten with 100% ethanol and then with 1x transfer gel equilibrium).
4. Put moisten 3MM-paper, fix the two face of blot paper with two clips on one top of another.
5. Blot-apparatus in coolroom, put the cover of the apparatus. Blotting: 250 mA- 45 min. (40V – 100W) or 0.8 mA / cm_2.Gel for 2.30 h (Fig. 6.4).

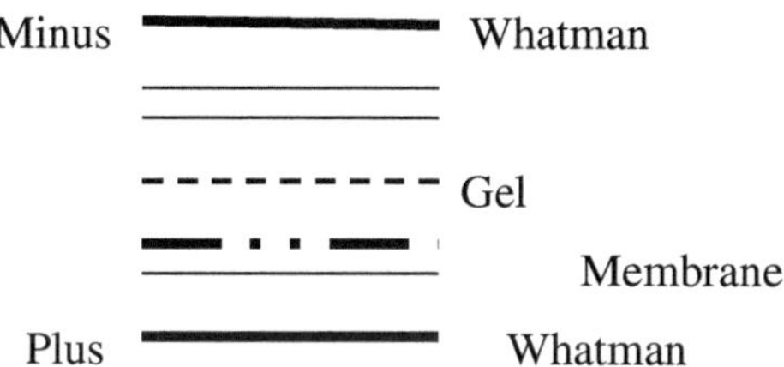

Fig. 6.4 Digrammatic representation of a Western blotting

6.10 Immune-Blot assay

1. Transfer protein from gel to a nitrocellulose membrane.
2. Place the membrane in a tray and block with 10 ml TBST buffer (5% Milch: TBST, 2.5 g: 50 ml TBST) for 1 hour at room temperature with shaking.
3. Incubate the blot with 10 ml of first antibody for 1 hour with shaking (1:100 dilution in blocking buffer).
4. Wash the membrane with 3 x 15 ml of TBST for 10 minutes each at room temperature.
5. Incubate with second antibody for 1 hour
6. Wash the membrane with 3 x 15 ml of TBST for 10 minutes each at room temperature.
7. Colour reaction: NBT + BCIP 1:100 dilution (colour development B., Fresh, reaction is stopped by adding water), and second antibody has alkaline phosphatase.

NBT stock solution: Prepare NBT stock solution by dissolving one tablet (yellow, round tablet) in deionized water, store at 4°C.

BCIP stock solution: Pepare BCIP stock solution by dissolving one BCIP tablet (25 mg substrate content, in 0.5 ml 100 % dimethylformamide), store at -20°C.

Substrate buffer: 0.1M Tris, 100 mM sodium chloride, 5 mM $MgCl_2$, pH 9.5, adjust pH with HCl.

Substrate solution: Prepare substrate solution bu adding 330 µl of NBT stock solution to 10 ml of substrate buffer, mix, then add 33 µl of BCIP stock solution, mix.

Nitro blue tetrazolium (NBT) is used in conjunction with the alkaline phosphatase substrate 5-bromo-4-chloro-3-indolyl phosphate (BCIP) in immunoblotting and immunohistological staining procedures.

6.11 Transgenic Organisms

In order to express foreign genes in plants or animals the coding sequences of the gene in question has to be inserted into an expression cassette. This cassette should provide promoter and terminator sequences. A number of techniques for manipulating DNA. One of these was the method for taking a piece of DNA from any source, and inserting it into a bacterial cell, in other words cloning it. The recipient bacteria may or may not express the protein the gene codes for, but the method of getting “foreign” DNA into a bacterial cell was essentially rather simple. Bacteria with the foreign DNA are said to have been transformed: their genome has been changed; new genetic characters have been inserted. Such bacteria may then be said to be transgenic, and it soon became the goal to make transgenic animals and plants. This was important as it would allow genes to be transferred across the species barrier in eukaryotes, a given gene from any organism. This works because the genetic code is universal.

6.11.1 Transgenic animal

Transgenic mice, and other animals, are usuall produced by rather tedious (but successful) method of microinjecting the DNA or gene into the eggs of the recipient organisms. It is not a brilliantly efficient method, but it is successful, although skilled operators are needed to carry out the microinjection. Typical shapes are as follows: out of every 100 eggs that are collected perhaps 85 are suitable for injection. Of those injected eggs placed survive the injection procedure, but only six of the injected eggs placed in the host mother result in live births, and only one or two of theses will be transgenic animals. Nevertheless, the procedure does work, and selected offspring are bred to produce a transgenic line of animals. There are a number of variables that at present are difficult to control. One is the dose of gene injected: several hundred copies of the gene may be injected, but an uncertain number of these are expressed. Furthermore, the injected DNA may be incorporated into the chromosomes of the host cell at random, may possibly not be stable, or may not br expressed. It may also be incorporated in the middle of one of the host own genes that is vital for life. Producing transgenic animals was a challenge, but why was it so important? First of all, it offered another way of finding out about eukaryotic genes and how they function. Transgenic mice, in addition to providing basic scientific knowledge about how genes function, may be very valuable as a means of studying human diseases. They have already been used in the investigation of cystic fibrosis, for example, through the production of mice containing the faulty human gene, and transgenic animals can give us deal of information about how genes work and they can provide "models" of human disease that give clues as to the cause of disease and might be used to test out potential treatments.

6.11.2 Transgenic plants (Plant gene transfer)

Foreign genes may be transferred into plant cells using viral vectors or by the direct transfer of DNA into protoplasts or intact cells, by particle bombardment.

Plants cells often have a close association with certain bacteria, e.g. *Agrobacterium tumefaciens*. This bacterium readily infects dicots (tobacco, carrot) and when it contains a plasmid (the so-called Ti plasmid) it will cause the formation of a plant tumour called a crown gall disease. This disease is characterized by the tumorous growth of plant tissues in the stem, and is a significant problem in the cultivation of grape vines, stone fruit and nut trees. *Agrobacterium*-induced the tumors were shown to be sources of auxin, and capable of growth in culture in the absence of both bacteria as well as the complement of plant growth regulators normally required to incite growth of callus from sterile plant tissues (T-DNA encoded genes for auxin and cytokinin biosynthesis responsible for tumor phenotype). Identification of tumor-inducing (Ti) plasmid narrowed the search to genetic elements derived from this plasmid and ultimately resulted in the discovery of T-DNA (transferred DNA), a specific segment transferred to plant cells. The tumors provides *Agrobacterium tumefaciens* with some advantage. This advantage derives from the production by tumors of unusual amino acid-like compounds called opines. Although opines are structurally diverse, a tumor produces only certain opines that are strictly dependent on the infecting strain, and the opines produced by a gall are specifically catabolized by that strain of *Agrobacterium*. Furthermore, the ability to metabolize opines is tightly correlated with virulence; loss of virulence is always accompanied by the loss of the ability to degrade a specific opine. Before the identification of the Ti-plasmid, these observations were the first indication that tumorigenesis involved the transfer of genetic material from bacteria to plants. It is now known that the enzymes for catabolism for specific opines are encoded on the Ti-plasmid and complement the opine biosynthetic pathways encoded on the T-DNA. Thus, by the introduction of genetic material into plant cells, `genetic colonization`. *Agrobacterium tumefaciens* creates a unique habitat wherein it solely is genetically equipped to utilize the predominant carbon nitrogen source. Here gene transfer is mediated by the Ti plasmid, in that the appropriate gene is placed between border sequences of the T-DNA and is stably integrated into the plant genome using the natural infection and insertion mechanisms of the bacterium. It is necessary to first clone the transgene into a binary vector, which can replicate in both *E. coli* and *A. Tumefaciens*, and contains *cis*-acting elements required for gene transfer into the plant genome.

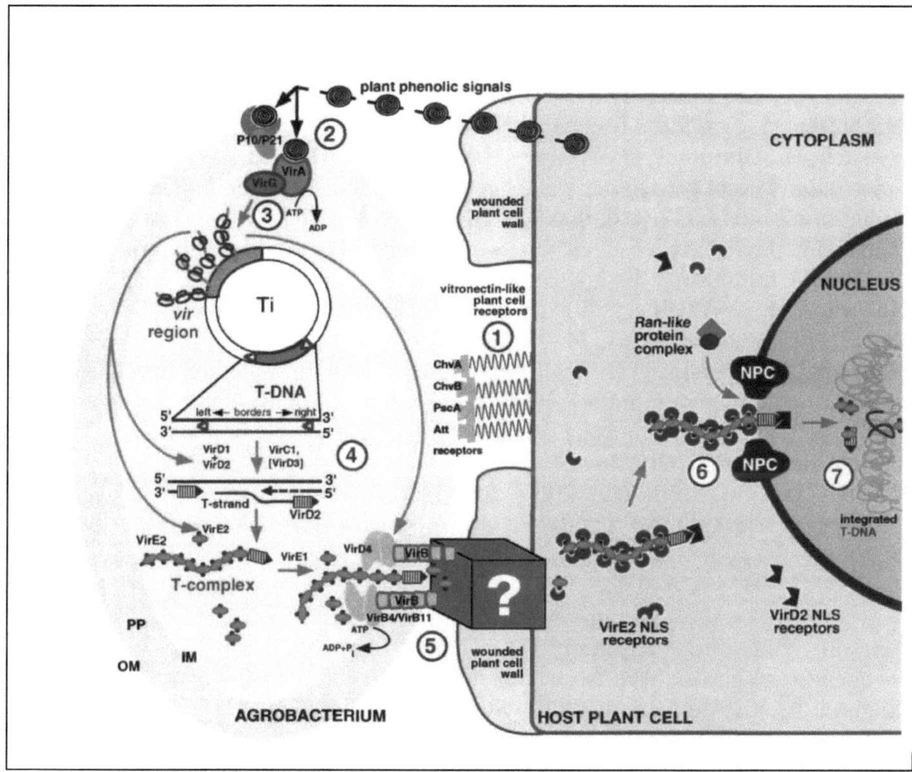

Fig. 6.1 Basic steps in the transformation of plant cells by *Agrobacterium tumefaciens* (see text for details). Adapted from Sheng and Citovsky (1996). **Where:** OM: outer membrane PG: Peptidoglycan cell wall; PP: periplasm; IM: Inner membrane; ◄-T-DNA border; NPC: Nuclear pore complex.

References

Abdel-Haliem, M.E.F (2004) Molecular-physiological analysis of two novel isoforms of phosphoinositide kinases from Arabidopsis thaliana (L.) Heynth. Ph.D. Max-Planck Institute of Molecular plant phys. Potsdam, Germany.

Allison, LA. (2007) Fundamental Molecular Biology. Blackwell publishing Ltd. Malden, MA 02148, USA.

Altschul, SF., Gish, W., Miller, W., Myers, EW., and Lipman, DJ. (1990) Basic local alignment search tool. J. Mol. Biol. 215: 403-410.

Brown, TA. (2003) Gene cloning and DNA analysis. 4th edn. Blackwell Science Ltd, Oxford.

Dressler, D., and Potter, H. (1982) Molecular mechanisms in genetic recombinant. Ann. Rev. Biochem. 51 : 727 -761.

Lehrach, H., Diamond, D., Wozney, J.M. and Boedtker, H. (1977) RNA molecular weight determinations by gel electrophoresis under denaturing conditions; a critical re-examination. Bioch. 16: 4743-4751.

Lipps, HJ., Jenke, ACW., Nehlsen, K., Scinteie, MF., Stehle, IM., and Bode, J. (2003) Chromosome-based vectors for gene therapy. *Gene* 304:23–33.

Maxam, AM., and Gilbert, W. (1977) A new method for sequencing DNA. Proc. Natl. Acad. Sci. USA. 74 (2): 560–564.

Moore, SJ., and Stein, WH. (1948) Photometric ninhydrin method for use in the chromatography of amino acids. J. Biol. Chem. 176:367-388.

Mullis, KB. (1990) The unusual origin of the polymerase chain reaction. Scientific American 262:36–43.

Nathans, D., and Smith, HO. (1975) Restriction endonucleases in the analysis and restructuring of dna molecules. Annual Review of Biochemistry 44: 273–293.

Niall, HD. (1973) Automated Edman degradation: the protein sequenator. Meth. Enzymol. Methods in Enzymology. 27: 942–1010.

Pearson, WR., and Lipman, DJ. (1988) Improved tools for biological sequence comparison. Proc. Nat. Acad. Sci. USA 85: 2444-2448.

Pingoud, A., and Jeltsch, A. (2001) Structure and function of type II restriction endonucleases. Nucleic Acids Research 29:3705–3727.

Rainer, D. (2001) Introduction to Molecular Biology. Blackwell Science, Inc. Malden, Massachusetts 02148, USA.

Sambrook, J., and Russell, DW. (2001) Molecular Cloning: a Laboratory Manual, 3rd edn. Cold Spring Harbor Laboratory Press, New York.

Sheng, J. and Citovsky, V. (1996) Agrobacterium-plant Cell Transport: Have Virulence Proteins, will Travel. The Plant Cell 8: 1699-1710.

Smith, HO. (1979) Nucleotide sequence specificity of restriction endonucleases. Science 205:455-462.

Sanger, F., and Coulson, AR. (1977) A rapid method for determining sequences in DNA by primed synthesis with DNA polymerase. J. Mol. Biol. 94: 441- 448.

Watson, JD. (2007) Recombinant DNA: genes and genomes: a short course. San Francisco: W.H. Freeman.

Printed by Books on Demand GmbH, Norderstedt / Germany